A Field Guide to the

# Amphibians and Reptiles of Madera Canyon

and adjacent areas

**John C. Murphy**

**Foreward by Daniel E. White**

ISBN 978-1-7378424-0-8


## Acknowledgments

Many people contributed to this book. My sincerest thanks go to all of them. Assistance in the field ranges from help finding animals to photography, to changing flat tires. But the goal was finding animals to photograph. I would like to thank Tom Anton, Alvin Braswell & Jan Weems; Roger Carter; Phil Collura; Joe Cavataio; Rene Clark; James Craig; Jason Fantuzzi; Larry L. C. Jones; Michael Maul; Tom Manos; Debbie & George Pohlmann; Kevin Pollack; Bryan Reynolds; Roger Repp; Larry Rowan, Mike Rutherford; Rhett Rautsaw, Jason Strickland, Andrew Mason, Jade Mellor, and Michael Buentempo.

For additional photographs, I sincerely thank Robin Agarval, Michael Andersen, Tom Anton, Rene Clark, Tim Cota, Lee Doolan, Antonio Gandini, Larry L. C. Jones, Yinan Li, Devon Medler, Susan Myers, Mason Salem, Diane Thornton, Laurie Vitt, Brad Wilson,

For comments on the content, writing, and photography my sincerest thanks go to Bob Pitcher, and Judy White for providing useful critiques, of all parts of the book.

A special thanks to Brett Blume, Director of the Santa Rita Experimental Range for access to the SRER

On the Cover. Face to face with a Tiger Rattlesnake. Photography by JCM

# FOMC - What we do.

The Friends of Madera Canyon is a U.S. Forest Service-affiliated non-profit volunteer organization having 501(c)(3) status. Informational brochures, cleanup teams, research and education programs for adults and schoolchildren, handicapped trail maintenance, trail and nature sign design and maintenance, visitor welcome, and funding for the construction of many facilities throughout the Canyon are all provided by the organization.

In response to the U.S. Forest Service's call for volunteers to help preserve and protect Madera Canyon, local businesses and residents formed the Friends of Madera Canyon in 1987 as the area's unique environment was threatened by environmental degradation and human encroachment. In recent years, fire has become a significant hazard.

The Friends of Madera Canyon support the educational, interpretative, scientific, and historical initiatives. To ensure that Madera Canyon continues to be one of Arizona's top natural destinations, our more than 500 household members (about 1,000 individual members) give 10,000 hours year on average.

We continue to confront new challenges. The canyon's natural splendor and the ecosystem's ecological integrity are constantly in danger. Proposed mining projects, new housing construction, and pumping water out of the aquifer could destroy the entire Santa Rita Sky Island environment. For additional information, get in touch with the Sky Island Alliance or Save the Scenic Santa Ritas.

You can join us at : https://friendsofmaderacanyon.org/join/

Contents

Yarrow's Spiny Lizard with mites (red areas behind ear opening).
Photography JCM

# Foreward

Education has been fundamental to the mission of the Friends of Madera Canyon since its founding in 1987. John C. Murphy's A Field Guide to the Amphibians and Reptiles of Madera Canyon is a major contribution to the resources now available to anyone seeking a road map for identifying, understanding, and appreciating a significant and numerous life form in the Canyon.

The Guide includes a brief introduction to the natural history of the Canyon, and a summary of how climate change has impacted, and will continue to impact, it. This book provides a snapshot in time of what amphibians and reptiles are here now as an historical record of potential use to future professional scholars and amateur herpetologists alike. The history of science is a sequence of such snapshots upon which researchers depend to understand how the past affects what exists now and what conditions might be in the future.

Aside from its value in the on-going work to understand the world in which we live, A Field Guide can also be useful to the hiker in the Canyon who is suddenly startled by a beautiful lizard or colorful snake and wonders what it is. Murphy's periodic "field notes"— "I have only seen this toad [Woodhouse] at Historic Canoa Ranch"— add personal observations in the way most good teachers do, connecting the more formal descriptions and discussions with "by the way" comments that engage the user of the guide more directly.

The photographs are noteworthy as well, collected by Murphy from a variety of sources, including his own pictures. Some of the toads seem to look directly into the lens with a face only a mother could love. Other plates, like the Eastern Collared Lizard, reveal patterns and colors on the skin that rival the most beautifully beaded jewelry. By themselves, the photographs are yet another addition to the historical record.

For me, field guides are a way to confirm, or not, what I think I saw, learn something new, and stimulate further inquiry about a species. The photos are useful and add depth and breadth of information, making the Guide an essential reference book. I cannot say that I always remember what I have learned from a guide. But I know where to look if I have forgotten.

John Murphy's work for A Field Guide was animated by his passion for the amphibian and reptile world. The number of books, articles and research reports he has published are testimony for his love of subject. He also loves Madera Canyon and currently serves on its Board of Directors. The Friends of Madera Canyon organization is pleased to benefit from his work and to offer a new and valuable resource to the wider community interested in the natural world.

**Daniel E. White, Board President**
**Friends of Madera Canyon**

It's never been more important for us to understand the effects of biodiversity loss… only if we do that, will we have any hope of averting disaster.

— Sir David Attenborough

From BBC TV program Extinction: The Facts (13 Sep 2020).

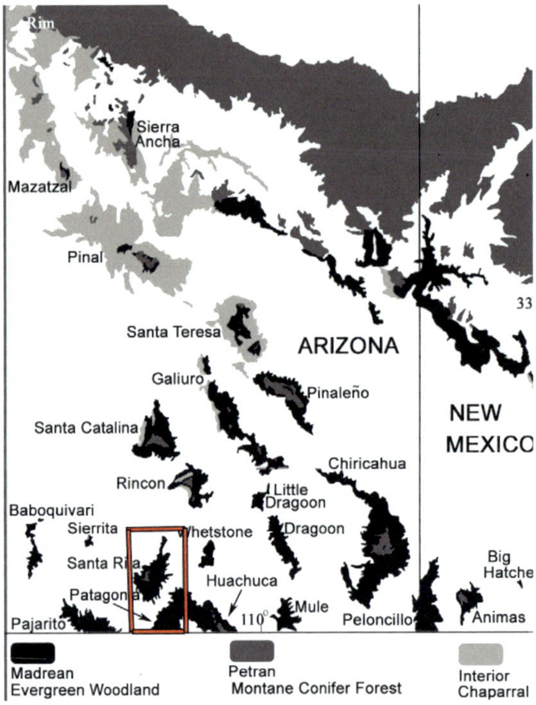


Figure 1. Southeast Arizona Sky Islands. The red rectangle marks the area discussed in this volume. The distribution of three montane biomes are shown in black and two shades of gray.

# Preface

Hundreds of thousands of people visit Madera Canyon, the Santa Rita Mountains, and the Patagonia Mountains each year. Many are local, but others come from other states and countries. Some come for a picnic or hike through the Madrean woodland, but many come to see the Elegant Trogon and the other birds that reach the most northern edge of their distribution in Arizona's sky islands. Yet others come to see frogs, lizards, and snakes. Or, they come to see butterflies and dragonflies and identify other insects or plants.

The geographical region discussed in this book are located in Pima and Santa Cruz counties in southeastern Arizona. This volume focuses on the Santa Rita Mountains, Madera Canyon, the Patagonia Mountains, and the intermontane desertscrub that is found around these ranges.

The Santa Rita and Patagonia Mountains are two of about 55 sky island complexes in Arizona, New Mexico, and Sonora, Mexico. In Arizona, the diverse landscapes, shaped by the convergence of the Colorado Plateau with the Mojave, Sonoran, and Chihuahuan deserts, produce an array of localized climates and microhabitats. The state exhibits a distinct variation in climate from lower elevations, characterized by desert conditions with mild winters and hot summers, to higher elevations, which are wetter with cold winters and mild summers. Notably, there are significant diurnal temperature fluctuations in desert and montane regions.

The state experiences rainfall in two distinct seasons. Winter rains, mainly influenced by Pacific cold fronts and prevailing westward winds, are cool and bring widespread, steady, and light precipitation. In contrast, late spring and summer have a shift in wind patterns from the south, marking the onset of the North American Monsoon (NAM). The monsoon is characterized by sporadic but intense storms fueled by moisture from the Gulf of California and the Gulf of Mexico. While the monsoon traditionally begins about July 4th and ends in September, it can commence earlier or extend into the fall, sometimes even into October or early November. If tropical cyclones from the eastern Pacific contribute moisture. This additional precipitation during the fall is known as the "chubasco."

The NAM was traditionally defined by consecutive days of high dewpoint. Recent definitions by the National Weather Service set its start on June 15th and end on September 30th. The actual dewpoint remains a more accurate indicator of the behavior of reptiles and amphibians.

Amphibians are remarkable for their adaptations throughout their life histories, which allow them to thrive in diverse ecological settings. Their reproduction is typically dependent on the availability of free water, making them less common in arid landscapes. However, many desert and thornscrub species synchronize their breeding with the NAM season, resulting in choruses of amphibians around stormwater ponds after the first heavy rains of summer.

This work provides species accounts for one salamander, 14 frogs, three turtles, 26 lizards, and 28 snakes that occur in Madera Canyon, the Santa Rita Experimental Range, the Santa Rita Mountains, the Patagonia Mountains, Canoa Hills, and adjacent areas (See Figure 1).

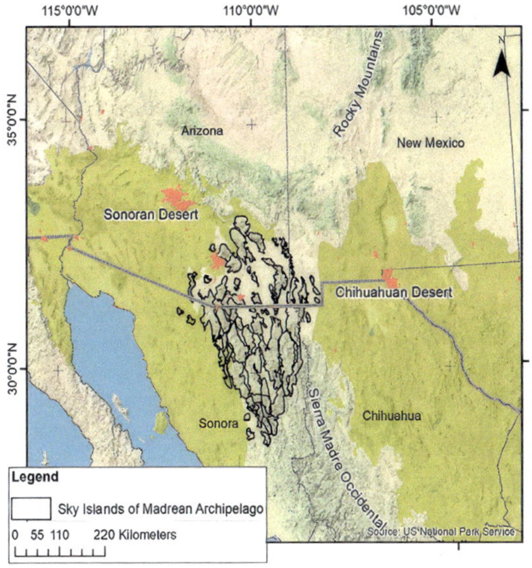


Figure 2. The position of the Madrean Sky Islands (black lines) links the Rocky Mountains and the Sierra Madre Occidental of Mexico. It connects the Chihuahuan Desert to the east and the Sonoran Desert to the west, making the area biologically diverse. From the National Park Service.

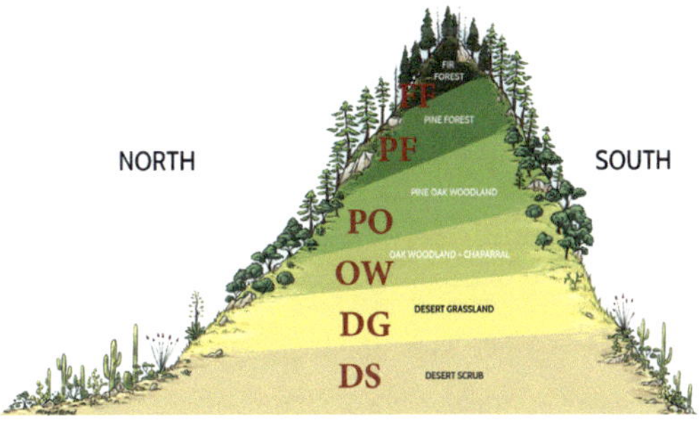


Figure 3. Plant biomes change with elevation, following a pattern similar to that from the equator to the poles. FF = Fir Forest, PF = Pine Forest, PO = Pine Oak Woodland, OW = Oak Woodland + Chaparral, DG = Desert Grassland, DS = Desert Scrub. Codes added. From Sky Island Alliance.

# Introduction

The Madrean Sky Islands are mountain ranges in northern Mexico and southeast Arizona. The mountains rise more than 6,000 feet above the surrounding desert and grasslands, creating a significant difference between the high forest-covered peaks and the low deserts and grasslands. The low-elevation habitats are inhospitable for many mid- and high-elevation species, and the reverse is also true for some species. By comparison, the deserts and grasslands are "seas," while the mountains are "islands" surrounded by them.

The US Forest Service manages the Sky Islands north of Mexico and they are part of the Coronado National Forest in southeast Arizona. Their biomes range from subalpine forests to subtropical deserts.

Drive I-19 south from Tucson, and the Santa Rita Mountain range is the first range you will encounter on your left. The Santa Ritas stretch over 25 miles from the Pantano Wash to Sonoita Creek, about 12 miles from the Mexican border. Most of the peaks and ridgelines in the Santa Ritas are between 6,000 and 7,000 feet above sea level, with Mount Wrightson, the highest peak, standing at 9,453 feet. To the east of the Santa Ritas is the San Pedro Valley, at about 4,500 feet. To the west is the Santa Cruz Valley, which has elevations of about 3,000 feet. The mountains act as rain collectors. Enough rain falls in the upper elevations to sustain forests. In contrast, at lower elevations to the west, the vegetation of the Sonoran Desert is prevalent, and to the east are grasslands. The Santa Rita topography is distinguished by its narrow ridges, jagged peaks, and steep slopes. The Mt. Wrightson Wilderness Area, also part of the Coronado National Forest, contains much of the Santa Rita and Patagonia ranges.

The Santa Rita and the Patagonia Mountains are connected by geology and shared plant and animal species. Two prominent peaks of the Patagonia Mountains are the 6,490-foot Guajalote Peak and Mount Washington, the highest peak at 7,221 feet.

This volume focuses on 2500 square kilometers of the Madrean Archipelago, encompassing the Santa Rita and Patagonia ranges and several outliers. These two ranges are separated by a mere ten miles of valley bottom at their nearest point. I will refer to this as the focus area shown in Figure 1.

The topography of the Sky Islands is defined by uplifted blocks of ancient crustal rocks forming the core or basement of the continental crust. The basin and range in southeastern

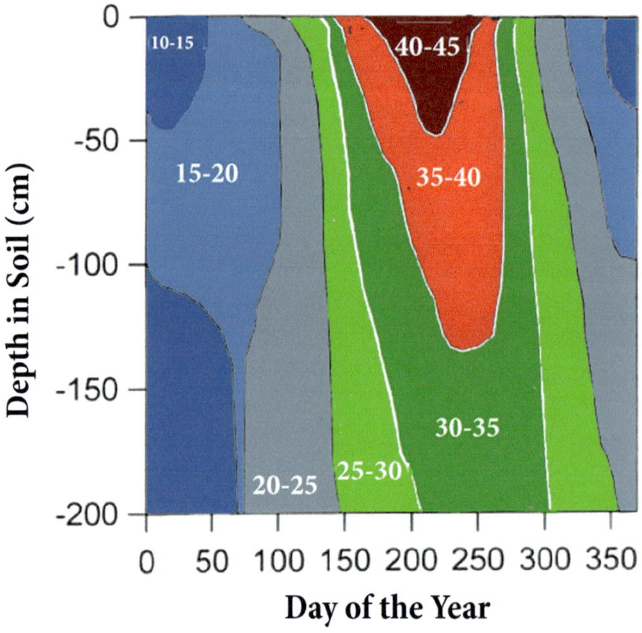


Figure 4 is adapted from Walsberg (2000) It represents the annual cycle of average daily soil temperature at a Sonoran Desert site. Daily minima and maxima vary greatly from average values in the upper levels of the soil. Data were collected in creosote bush scrub north of Gila Bend, Maricopa County, Arizona. Data were collected at 0, 10, 20, 30, 40, 60, 80, 100, 125, 150, and 200 cm below the soil surface using thermocouples. Values were collected over 24 hours at 4-week intervals throughout 1998 and then interpolated to form best-fit isotherms. Here are critical thermal maximums for some reptiles: Sonoran Desert Tortoise (39.5-43°C), Glossy Snake (41-42.5°C), California Kingsnake (42°C), Coachwhip (42.4°C), Spotted Leaf-Nosed Snake (38-39.3°C), Gopher Snake (40.5°C), Leaf-Nosed Snake (41.5°C), Western Diamondback (39°C), Twin Spotted Rattlesnake (42°C), Gila Monster (42.5-44.2 °C), Sideblotch Lizard (48.4°C), Collard Lizard (46.5°C), Desert Spiny Lizard (43.9°C) . See text for discussion. Critical thermal maximum data from Brattstrom (1965). Critical thermal maximum it's the temperature at which an animal loses its ability to escape from conditions that will lead to its death.

Arizona took shape between thirty-five and six million years ago (Scarborough and Brusca 2015). Since then, numerous plant and animal species have colonized the bajadas, foothills, and mountain ranges. Some species have become extinction, while others have adapted and evolved to survive the tests of time.

Bajadas are found at low elevations around the Sky Island ranges and are formed by eroding rock particles that form alluvial fans. Those fans then join to form the bajada. Thus, a bajada consists of coalescing alluvial fans along a mountainfront. These fan-shaped deposits are formed by depositing sediment within a stream onto flat land at the base of a mountain.

Throughout the Pleistocene advancing and retreating glaciers played a pivotal role in shaping the distribution of many species. These glacial cycles intensified in the last million years, causing shifts in latitudinal and elevational ranges. During previous glacial periods, sky island habitats descended to lower elevations, enabling formerly isolated populations to come into contact at mid-to-high elevations. Landscape transformations have proven to be significant catalysts for evolutionary processes.

In a 1943 article titled *Monument in the Mountain*, published in Arizona Highways magazine, Natt N. Dodge eloquently referred to the Chiricahua Mountains in southeastern Arizona as a "mountain island in a desert sea." This notion of terrestrial islands, encircled not by water but by low-elevation terrestrial landscapes, was also explored by Weldon F. Heald in his 1967 book, *The Chiricahua Mountains*. Heald, who served as the director of the Sierra Club in the early 1960s and as a consultant on National Parks and National Monuments to Stewart Udall, Secretary of the Interior, recounted his firsthand experiences in the Chiricahua Mountains while expanding the concept of sky islands.

The term "Madrean Archipelago" was first coined by Charles H. Lowe in 1992 to describe the Sky Island ranges extending from the Sierra Madre Occidental in Sonora and Chihuahua to the Mogollon Rim in central Arizona (Van Devender et al. 2013). Subsequently, Warshall (1995) and McLaughlin (1995) further refined and expanded the geographical area, solidifying the concept of the Madrean Archipelago and its unique ecological significance.

The Santa Rita Mountains encompass seven distinct biomes. These biomes include Desertscrub, Semidesert Grassland, Interior Chaparral, Madrean Encinal, Madrean Evergreen Woodland, Mixed Coniferous Forests, and Riparian Vegetation. The classification of these biomes in the Santa Rita Mountains is based on the work of Brown (1994). Each of these biomes represents a unique combination of plant and animal communities adapted to the specific environmental conditions found in the region.

## Other Herpetological Studies in the Focus Area

Previous studies of amphibians and reptiles have received the least amount of attention by ecologists in the focus area. Much of the research in the area was done on the Santa Rita Experiermental Range (SRER).

Studies of amphibians in the focus area have included conservation efforts to re-establish populations of the Tarahumara Frog (*Lithobates tarahumarae*) (Rorabaugh et al. 2005) and to report the discovery of species previously not known to occur in the area, such as the Barking Frog (Rorabaugh 2004).

A study on the distribution of rattlesnakes was based on 40 records of Western Diamondbacks (*Crotalus atrox*), six records of Tiger Rattlesnakes (*C. tigris*), seven records of Mohave Rattlesnakes (*C. scutulatus*), and nine records of Blacktailed Rattlesnakes (*C. molossus*). Diamondbacks ranged from an elevation of 854 to 1,220 m. Mohave rattlesnakes ranged from an elevation of 854 to 1,373 m, and Blacktails were found in canyons from 1,281 to 1,464 m. The distribution of Tiger Rattlesnakes overlapped the distribution of all the other rattlesnakes in the study (Humphrey 1936).

Echternacht (1967) studied whiptail lizards (*Aspidoscelis exsanguis = Aspidoscelis sticto-grammus* and *A. tigris gracilis*) on the SRER for diet overlap.

As mesquite was cleared from the SRER in various treatments, the Sonora Whiptail (*Aspidoscelis sonorae*) became more abundant than in areas that contained undisturbed mesquite and mesquite with irregularly shaped clearings (Germano and Hungerford 1981). Germano and Hungerford (1981) studies were pioneer studies in considering reptiles in landscape management plans in the Southwest.

# Conversion Table

Table 1. Conversion table for temperature, length, and weight from Metric to English.

| Metric value | Multiplied by | English values in |
|---|---|---|
| Celsius | (°C x 1.8) + 32 | Fahrenheit |
| Centimeters (cm) | 0.39 | Inches |
| Millimeters (mm) | 0.039 | Inches |
| Meters (m) | 3.28 | feet |
| Kilograms (kg) | 2.20 | pounds |

# Threats

A Canadian mining firm, HudBay is threatening the ecological integrity of the Santa Ritas. Hudbay's Copper World is not yet in production. Save the Scenic Santa Ritas (SSSR) has contested the air permit, but Copper World has been given its water and air permits by the state. The SSSR has a case ongoing to withdraw the right-of-way, but they require it to execute their plans to move trucks across the Santa Rita Experimental Range.

Hudbay is currently gathering data to prepare a final feasibility study and seeking a joint venture partner. Should they proceed, they will need final approval from their board of directors.

The Copper World project will dig and blast its way to 4 or 5 open-pit mines, and it will build a plant to manufacture sulfuric acid on-site. Excess production of sulfuric acid will be sold to other companies, which means that they will truck the excess acid off the mountain and sell it to buyers in the USA and Mexico.

Copper World will pollute the air and water table and create dust pollution from trucks driving across the bajada. They will draw down the water table and compete for water with the agricultural and human populations. Copper World will harm the water supply while competing with eco-agriculture and humans for water resources.

While society needs copper - it is not to the degree mining companies suggests. The mines will be in the Madrean Biodiversity hotspot.

A public land sale also threatens Madera Canyon and the Santa Ritas. The land sale is a give-a-way to wealthy it threatens the biodiversity with habitat fragmentation, urban sprawl, and agricultural development.

Then there is climate change.

## Climate Change

Amphibians and reptiles rely on sunlight, air, water, and substrate temperatures for-life-supporting physiological processes, including muscle contraction, food digestion, and-cues for the timing of reproductive events. Reptiles have a thermal niche, which may be altered by a change in temperature regimes and decrease their survival ability. Amphibians are thermal conformers and can be active at cooler temperatures than reptiles. Climate change can lead to food scarcity, courtship failures, and reproductive failures. Lizard mortality may be linked to warm spells in winter, vegetation changes, fire regimes, and invasive species. Snakes and lizards face new diseases and may lose the ability to cope with old and new pathogens when exposed to temperature extremes. Turtles may find their habitats fragmented due to an altered climate, and temperature-sensitive sex determination may affect their populations structure. Altered thermal niches can lead to lethal conditions, such as habitat damage due to extreme weather.

The impact of climate change on amphibians and reptiles is significant, and will continue to be significant throught the next century.

## Abbreviations Used

ASL = above sea level
C = degrees Celsius
cm = centimeters
(i) = introduced
kHz = kilohertz
(a unit of measurement that represents 1,000 cycles per second)
km = kilometers
m = meters
mm = millimeters
MYA = millions of years ago
p/s = pulses per second
SUL = body length of frogs and toads
SVL= body length of salamanders, lizards, and snakes

## Conservation Status Abbreviations

The conservation status given for each species or subspecies by Nature Serve is located on the line for the scientific name of the taxa. Abbreviations used are: (S) = secure, (AS) = apparently secure, (VU) = vulnerable, (IM) = imperiled, (CI) = critically imperiled, (nd) = no data.

Note that species accounts that have a question mark (?) in front of the Figure numbers are species considered problematic in the study area but the author has reason to suspect their presence.

Also note that parenthesis around the author(s) names for a given species means that species is now in a different genus than when it was originally described.

Appendices 1-5 contain labeled illustrations of Amphibians and Reptiles that will help explain terminology and identify species.

The Glossary is a list of terms frequently found in herpetological papers and books, including this one.

## The Santa Rita Experimental Range

The Santa Rita Experimental Range (SRER) is the longest continuously active rangeland research facility and among the five oldest biological field stations in the United States. Located south of Tucson in Pima County, Arizona, the 52,000-acre range was founded in 1902 and administered by the United States Forest Service until 1987, when the University of Arizona College of Agriculture took over the administration of the site. It is currently part of the University of Arizona, Arizona Experiment Station system.

The SRER is a world-class facility because of the long-term historical and biological databases that have been maintained since its creation. Indeed, it has been a principal site for pioneer range research with the purpose to build a sound body of scientific knowledge on the restoration, protection, and management of semiarid grasslands in the arid Southwest. The accumulated information on the ecology of the semidesert system at SRER is more complete than for any other tract of comparable size and diversity. Results of this research have now direct applicability to over 20 million acres of semiarid rangelands in the United States and to another 20 million acres in northern Mexico.

# Amphibians

Amphibians are found on all land masses, except in the polar regions of Antarctica and Greenland. The highest diversity of amphibians can be found in the neotropics. They show a diversity of life histories with different modes of reproduction that are spectacular. The word "Amphibian" comes from the Greek words "amphi-" and "bios," which refers to their two-stage lives - aquatic and terrestrial. They are ectothermic and thermal conformers, meaning their body temperature is determined by their environmental temperature. Some have evolved proteins that help them survive freezing temperatures. Amphibians have permeable, moist, and glandular skin that absorbs oxygen and environmental cues. The most prominent family of salamanders, the Plethodontidae, don't have lungs and use their skin for exchanging gases with the atmosphere or water. Amphibian skin glands also produce toxins, some of which can be fatal to other species. However, some of these toxins have medical and pharmaceutical uses.

Unfortunately, amphibians face declining populations due to climate change, diseases, and habitat destruction. Living amphibians are frogs and toads (Anura), salamanders and newts (Caudata), and the tropical, wormlike caecilians (Apoda).

# Salamanders & Newts

There are approximately 200 species of salamanders in the US, and their distribution is noticeably skewed toward the northern hemisphere. The three primary amphibian groups—Apoda (caecilians), Urodeles (salamanders), and Anurans (frogs)—remain the subject of intense discussion among paleontologists and herpetologists on their origins and evolutionary relationships.

Although their life cycles differ significantly, salamanders usually lay their eggs in water and produce aquatic larvae. In several settings, certain species reproduce while they are still larvae. Salamanders possess the ability to regenerate injured or lost body parts. What makes this so amazing is that it happens without leaving any kind of scar. Because of this, salamanders are model organisms in scientific studies that strive to comprehend and realize regenerative processes for advances in human and animal biology and medicine.

# Mole Salamanders

The Mole Salamander family Ambystomatidae contains two genera, *Ambystoma* (the mole salamanders) and *Dicamptodon* (the Pacific Giant Salamanders). *Ambystoma* contains 32 species widely distributed across North America, while *Dicamptodon* contains four species restricted to the Pacific Northwest. They are primarily terrestrial and feed on invertebrates and small vertebrates. Some species retain juvenile traits as mature adults. Arizona has three subspecies of Western Tiger Salamanders: *Ambystoma m. stebbensi*, often known as the Sonoran Tiger Salamanders; *A. m. nebulosum*, the Arizona Barred Salamanders; and *A. m. mavortium,* the Barred Tiger Salamanders. In the Amphibians of the World account of *Ambystoma mavortium*, Frost commented that in the Western USA if a bass fishery is present the water will contain a population of tiger salamanders derived from populations scattered around the USA [https://amphibiansoftheworld.amnh.org/Amphibia/Caudata/Ambystomatidae/Ambystoma/Ambystoma-mavortium].

## Western Tiger Salamander

*Ambystoma mavortium* Baird, 1850 (CI) Figures 5-6.

**IDENTIFICATION**. The Western Tiger Salamander is among the largest terrestrial salamanders in North America, with adults exceeding 35 cm in total length. It distinctive appearance is characterized by a black dorsum, a rounded snout, and slightly bulging eyes. The tail is laterally compressed, which aids in swimming.

**DISTRIBUTION.** This salamander is primarily found in the Patagonia Mountains and San Rafael Valley of Arizona, while its presence in the Santa Rita Mountains remains uncertain. Within Arizona, you'll typically encounter the Sonora Tiger Salamander. The species' distribution in Mexico is less clear, though *A. m. stebbinsi* has been reported in stock tanks in northern Sonora.

Figure 5. An adult Western Tiger Salamander, *Ambystoma mavortium*. JCM

Figure 6. A Western Tiger Salamander adult that has sexually matured, lost its juvenile traits, and become terrestrial. Photography Antonio Gandini.

**HABITATS.** *A. mavortium* habitats have become artificial, with cattle tanks serving as substitutes for natural bodies of water that were once prevalent in wetland areas like the San Rafael Valley. Aquatic habitats are crucial for reproduction, typically required from January to June. Adult terrestrial salamanders are often found in grasslands and oak-juniper forests, where they use loose soil and mammal burrows for protection against harsh weather.

**REPRODUCTION.** Breeding occurs after winter rains, prompting adults to migrate from terrestrial sites to breeding ponds. The breeding season usually begins in January and can last until early May, with females laying between 200 and 2,000 eggs affixed to various substrates. Eggs hatch in 2 to 4 weeks, and the larvae, initially feeding on zooplankton, gradually shift to larger dietary items as they grow. Interestingly, some young salamanders may undergo metamorphosis within just two months, transitioning to a terrestrial lifestyle, while others that develop in permanent water may remain aquatic as neotenic adults, retaining their gills and aquatic traits throughout their lives.

**LARVAL FEEDING BEHAVIOR.** Larval salamanders primarily consume zooplankton, but as they mature, they shift to larger macroinvertebrates. Adult terrestrial salamanders feed on insects and sometimes on the eggs and larvae of their species during the breeding season.

**CONSERVATION STATUS.** The Western Tiger Salamander, *Ambystoma mavortium*, was first described by Baird in 1850. This species is generally considered secure; however, its subspecies, the Sonora Tiger Salamander (*A. m. stebbinsi)*, is listed as Critically Imperiled (CI) and is fully protected in Arizona.

The Western Tiger Salamander faces several threats, including habitat loss due to drought, flooding, and the degradation of cattle tanks. Urban development and infrastructure fragmentation further threaten their populations. Protecting their aquatic habitats, especially during critical breeding times, is paramount for the survival of this species.

**TAXONOMIC NOTE..** Three subspecies of the Western Tiger Salamander *A. mavortium* are found in Arizona. The Barred Tiger Salamander, *Ambystoma mavortium mavortium*, is indistinguishable from the Arizona Tiger Salamander, *Ambystoma mavortium nebulosum.* These two subspecies are likely the result of inbreeding between *Ambystoma mavortium mavortium* and *Ambystoma mavortium nebulosum.* Arizona *A. m. mavortium* are invasives brought to Arizona by the fishing bait industry. The Sonora Tiger Salamander, *Ambystoma mavortium stebbinsi,* is endangered by the USFW and is restricted to southeastern Pima County and Santa Cruz County usually above 5000 ft. *Ambystoma mavortium* has had a long, confusing taxonomic history. It has been described twelve times and has at least 22 scientific name combinations.

The origin of the three subspecies of the Western Tiger Salamander, *Ambystoma mavortium*, in Arizona, involves a complex interplay of historical biogeography, hybridization, and environmental adaptation. The subspecies include the barred tiger salamander (*Ambystoma mavortium*), the Arizona tiger salamander (*Ambystoma mavortium nebulosum*), and the Sonoran tiger salamander (*Ambystoma mavortium sonoriense*). Understanding their origins requires examining genetic divergence, historical introductions, and ecological factors influencing their distribution.

Genetic studies indicate significant divergence among the subspecies of **Ambystoma tigrinum**, with high molecular divergence estimates between eastern and western forms, suggesting a long history of separation and adaptation to different environments. The introduction of non-native barred tiger salamanders into California by bait dealers has led to hybridization with native populations, raising concerns about genetic integrity and conservation 2,3. This hybridization has been documented to occur over several decades, creating a complex genetic landscape where native and introduced alleles intermingle 4,5. The introduction of *A. tigrinum mavortium* into Arizona, particularly in the Sulphur Springs Valley, has been linked to similar patterns of hybridization and genetic introgression, impacting local populations of the native Arizona tiger salamander.

Ecologically, the Western Tiger Salamanders exhibit a range of life history strategies influenced by environmental conditions. For instance, the presence of ephemeral ponds, which are critical for breeding, has been shown to affect the survival and growth of both native and hybrid salamanders The adaptability of these salamanders to varying hydroperiods and habitat types is crucial for their persistence in changing environments, particularly in the context of climate change. Furthermore, the fitness trade-offs associated with different life history morphs, such as paedomorphs and metamorphs, highlight the role of environmental variation in shaping the evolutionary trajectories of these subspecies.

In summary, the origins of the three Western Tiger Salamander subspecies in Arizona are rooted in a combination of historical biogeography, anthropogenic influences through introductions and hybridization, and ecological adaptations to their environments. The ongoing interactions between these factors continue to shape the genetic and ecological dynamics of *Ambystoma mavortium* in the region.

# Frogs & Toads

Frogs are a diverse and primarily carnivorous group of short-bodied, tailless amphibians composing Anura. They are widely distributed, ranging from the tropics to subarctic regions, with the greatest concentration of species diversity in tropical rainforests. Frogs account for around 88% of extant amphibian species and are among the five most diverse vertebrate orders. Adult frogs have stout bodies, protruding eyes, an anteriorly attached tongue, limbs folded underneath, and no tail. They have glandular skin with secretions ranging from distasteful to toxic and live in habitats ranging from freshwater to dry land. Frogs have a diet mostly consisting of small invertebrates. Tadpoles sometimes feed on plant matter, detritus, bacteria, and algae. Frogs have a rich microbiome and efficiently con vert their food into body mass. Frog populations have declined significantly since the 1950s, with over one-third of species threatened with extinction and more than 120 extinct since the 1980s. Conservation biologists are working to understand and resolve these problems.

# American Spadefoots

The American Spadefoots, family Scaphiopodidae, are burrowing frogs with short legs and protruding eyes. They have a stiff, keratin spade on their feet for digging. Like most fossorial frogs, they dig backward, scooping out the soil below their body. They differ from toads (Bufonidae) in having vertical pupils, while toads have round pupils. The shape of the spade can distinguish the two genera. *Scaphiopus* has a cycle-shaped spade, and *Spea* has a wedged-shaped spade.

---

## Couch's Spadefoot

*Scaphiopus couchii* Baird, 1854 (SE). Figures 7-8.

**IDENTIFICATION.** Couch's Spadefoot is a striking amphibian that can reach up to 90 mm in snout-vent length (SUL). In Arizona, males typically measure between 62 and 74 mm, with those in amplexus averaging 51 to 75 mm. Their dorsal coloration ranges from green to yellow-green, with large eyes featuring vertical pupils. A distinct dark, sickle-shaped tubercle is present on the ventral surface of each hind foot, which is longer than wide. Notably, there are no cranial crests or parotoid glands. Females exhibit dark reticulations, while males, which are slightly smaller, tend to be lighter with minimal dark markings. This differentiates them from other Arizona spadefoots, which are typically brown, tan, grey, or grey-green and lack reticulations, sporting keratinized, wedge-shaped spades on their hind feet instead of the sickle-shaped tubercles seen in Couch's Spadefoot.

**VOCALIZATION.** Couch's Spadefoot is known for its distinctive advertisement call that can be described as a "loud resonant yē-ŏw" repeated at intervals, resembling a plaintive bellow or the anxious bleating of a sheep. The call lasts between 0.75 and 1.25 seconds, with intervals of 3 to 5 seconds between calls.

**TADPOLE.** The tadpoles of Couch's Spadefoot are small, measuring 18 to 24 mm in total length, and are characterized by a black coloration with iridophores dispersed evenly over their bodies and tails. Newly transformed spadefoots are about 9.5 to 12.9 mm long, with females displaying mottled dark reticulations, while males have a more uniform color. Their dorsum ranges from dark brown to black with mottling against a yellow-to-green background, complemented by a bright white belly and black-brown irises with bronze flecks.

**REPRODUCTION.** Reproductive activity occurs during a pelvic amplexus, with males clasping females around the waist as they release eggs and sperm simultaneously. A female can lay more than 3,000 eggs per clutch, typically deposited on vegetation or directly into the water if plants are lacking. The embryos develop rapidly, with tadpoles potentially emerging in less than 24 hours, feeding primarily on detritus. Successful reproduction requires ephemeral pools that persist for at least seven to eight days; pools that dry up prematurely can lead to unsuccessful breeding. Longer-lasting pools produce larger larvae and metamorphs, emphasizing the rapid developmental speed of this species.

**HABITAT**. Couch's Spadefoot thrives in the Chihuahuan and Sonoran Deserts across various landscapes, including short-grass prairies, mesquite savannas, creosote flats, and semi-desert grasslands. They often inhabit alluvial fans near mountain ranges and dunes where the soil is conducive to burrowing. The species adapts to clay soils that can retain water in temporary ponds for at least a week. Burrows may be excavated by the frogs themselves or borrowed from mammals, providing both protection and moisture.

**FEEDING BEHAVIOR.** Newly transformed froglets primarily consume arachnids and insects, especially beetles, springtails, flies, and wasps. Adults have a diet rich in winged and nymph termites and consume a significant variety of other arthropods, being able to eat up to 55% of their body weight, which provides energy reserves lasting about a year. They typically avoid arthropods with known chemical defenses.

**DEFENSE.** Couch's Spadefoot has developed a defense mechanism like other amphibians, producing noxious substances in their skin. This secretion can cause irritation, resulting in symptoms such as a runny nose and sneezing in humans and likely affect other moist membranes.

**FIELD NOTES**. To observe Couch's Spadefoot, the best time is during or after a monsoon rain. They are frequently seen on blacktopped roads, and all age classes, including adults in amplexus, can be spotted in these conditions. Evening walks in the desert shortly before or after rainfall may offer glimpses of their active feeding behavior.

Figure 7. A male Couch's Spadefoot calling in a flooded ditch along Whitehouse Canyon Road. Note the leech attached to the spadfoot's knee. JCM

Figure 8. A pair of Couch's Spadefoots in amplexus. JCM

## Chihuahuan Desert Spadefoot

*Spea multiplicata stagnalis* (Cope, 1875) (SE) Figures 9-10.

**IDENTIFICATION.** The Chihuahuan Desert Spadefoot is a relatively small amphibian, with adults typically reaching 65 mm in snout-vent length (SUL). This species is distinguished by red-tipped tubercles scattered across its dorsum, which is uniformly brown or dark grey, featuring small dark spots or larger blotches. Unlike some other spadefoot species, it lacks the bony boss between the eyes; the distance between the eyes is greater than their diameter. Each heel has a short, wedge-shaped spade, and the iris is variegated, exhibiting a pale copper hue. Males possess heavily pigmented vocal sacs.

**DISTRIBUTION.** The Chihuahuan Desert Spadefoot (*Spea multiplicata stagnalis*) is primarily found in the eastern half of Arizona, inhabiting a variety of arid and semi-arid environments, provided that breeding pools are accessible. In Arizona, these spadefoots spend most of the year underground, emerging with the onset of monsoon rains in July and retreating back into their burrows by September.

Figure 9-10. Two Mexican Spadefoots show some of the color varition in this species. JCM

**HABITAT**. They are found in various deserts and grasslands, including Chihuahuan Desert Scrub and Great Basin Grasslands. It thrives in habitats such as sandy or gravel soils within semi-desert grasslands, sagebrush flats, semi-arid shrublands, river valleys, and agricultural areas.

**DIET.** As a generalized arthropod predator, the Chihuahuan Desert Spadefoot primarily consumes insects and spiders, which make up over 90% of its diet. Studies of New Mexico specimens in playas indicate that Carab beetles and weevils (Curculionidae) are the most frequently consumed food items. While this species generally avoids arthropods with noxious chemical defenses, like blister beetles, velvet ants, stink bugs, and millipedes, it may occasionally feed on centipedes and scorpions. Research has shown that individuals may require seven feedings to accumulate sufficient fat reserves for survival over a year.

**TADPOLES.** Larvae exhibit two known morphs: carnivorous and omnivorous. The carnivorous morph specializes in feeding on fairy shrimp (*Anostraca* sp.), featuring a broad head and large jaw muscles, developed for rapid growth. Conversely, the omnivorous morph feeds on detritus, algae, and occasional fairy shrimp. There is a notable trade-off in these strategies, with carnivorous morphs developing more rapidly in shrinking pools, while omnivorous morphs tend to have a survival advantage post-metamorphosis in ponds that remain filled with water.

**HYBRIDIZATION**. The Chihuahuan Desert Spadefoot can hybridize with the Plains Spadefoot (*Spea bombifrons*). Hybrid offspring tend to develop faster than individuals from either parent species, making hybridization adaptable in environments where ponds may dry up before full metamorphosis can occur.

**DEFENSE.** This species possesses the ability to exude noxious secretions that can become airborne, releasing an odor reminiscent of pungent peanuts. These secretions can act as a strong irritant to humans and potential predators.

**FIELD NOTES.** Chihuahuan Desert Spadefoots are less frequently observed compared to Couch's Spadefoot, often requiring diligent searching. In a population of Couch's Spadefoots, it's necessary to examine 20 to 30 individuals before encountering a Chihuahuan Desert Spadefoot. Their activity is closely tied to weather patterns, and they are best seen following rains, particularly during their brief periods of emergence.

# Toads

Members of the family Bufonidae are commonly referred to as toads. They have a worldwide distribution with about 52 genera and 600 species. The family is absent from Australia and the islands of the South Pacific, except for the introduced Neotropical Cane Toad, *Rhinella marina.*

Toads usually have short limbs for walking or hopping and dry, warty skin. Most have enlarged poison-producing parotoid glands behind the eyes. This clade has a variety of reproductive strategies. All Arizona species lay eggs in water and produce aquatic larvae.

See Appendix 2 for spadefoot and toad comparisons.

---

## Great Plains Toad

*Anaxyrus cognatus* (Say in James 1822) (SE) Figures 11-12.

**IDENTIFICATION.** Great Plains Toads are medium to large toads. Males typically measure between 47–103 mm in body length, while females range from 49–115 mm. In Arizona, males average about 68.8 mm and females about 72.8 mm. Great Plains Toads are easily recognizable due to their distinct patterns of dark green spots outlined in white, along with large blotches on their backs. Their skin color can be grey, brown, or green, with darker spots on top. The belly is usually a uniform cream or white with a yellow or orange patch. Some toads may have a thin, light stripe running down their back.

The skin of adult Great Plains Toads has many small warts, and they have noticeable crests on their heads , and a lump on the snout. Their parotoid glands, which are elongated, are located behind the eyes. They have a symmetrical pattern of large, light-edged dark spots on their backs.

**VOCALIZATION.** Their call sounds like a loud jackhammer and can last anywhere from 5 to 50 seconds.

**REPRODUCTION.** Heavy rains cause these toads to come out of their burrows and move to wet areas for breeding. In Arizona, they typically reach reproductive maturity in their second year. They breed explosively, meaning they breed in large numbers for a short time. Females may lay eggs multiple times during the active season. Males often gather in groups of 200–500 to call for females, but smaller groups of six to ten are more common. Males usually position themselves along vegetated edges of ponds.

The eggs are black on top and white underneath, about 1.2 to 1.3 millimeters in size. They are laid in long strings, usually in a single layer.

**LARVAE.** Tadpoles start off black with light flecks, then become lighter and mottled

Figure 11. A male Great Basin Toad calls from a temporary pond. Photography JCM

Figure 12. A Great Plains Toad. Photography JCM

brown as they grow. They have a highly arched dorsal fin and can measure between 25 to 35 mm in length. Tadpoles in permanent ponds face more predators than those in temporary ponds. Tadpoles take 17–45 days to metamorphose. After turning into toads, they stay near their birth pond for about a month before venturing into the surrounding area. These newly transformed toads often gather together.

**HABITAT.** Great Plains Toads live in a variety of habitats, including prairies, sandhills, and deserts. They can tolerate agricultural areas and drier conditions better than many other toad species.

**FEEDING BEHAVIOR-DIET** These toads hunt mostly at night, feeding primarily on insects. A population in New Mexico mainly eats flies, moths, beetles, and other small arthropods.

**PREDATORS.** Adult toads have predators like badgers, skunks, and crows, while tadpoles are eaten by birds and other aquatic creatures. Great Plains Toads dig shallow burrows for short breaks, but during extreme weather, they dig deeper (up to 55 cm), often creating a question mark shape with the toad at the bottom.

**REPRODUCTION.** On average, a female lays around 11,074 eggs, but this can range from 1,342 to 45,054. The number of eggs increases with the size of the female, and females can lay multiple clutches each season.

**FIELD NOTES.** Great Plains Toads are not commonly seen. I have found small groups calling and individual toads looking for food. The best way to locate them is driving on rainy nights and listening for their calls.

## Red-Spotted Toad

*Anaxyrus punctatus* (Baird & Girard, 1852) (SE) Figure 13.

**IDENTIFICATION.** Red-spotted Toads are small and attractive, with adults reaching up to 76 mm in body length. Females are generally larger, averaging 59 mm, while males average 52 mm. Tadpoles can grow to about 30-32 mm, and newly transformed toads are around 13-18 mm long. They have a flattened head and a body that ranges from brown to grey, adorned with small red spots that may be less visible in older individuals. Their parotoid glands are small and round, and the belly is usually white or cream, sometimes with black spots in younger toads. Unlike many other toad species, Red-spotted Toads do not have prominent crests on their heads.

**LARVAE.** The tadpoles are mostly black with bronze flecks and have a transparent tail, often found in large groups in muddy stream habitats. They typically remain in the larval stage for about eight weeks in central Arizona.

**VOCALIZATION.** Males have a single vocal pouch under their throat, and their call is often described as high-pitched and bird-like.

Figure 13. An adult Red-spotted Toad. Photography JCM

**REPRODUCTION**. Breeding occurs from March to September, depending on the habitat. In the spring, they use small streams, while in the summer, they breed in temporary pools created by monsoon rains. Adults usually stay near these water sources throughout the year.

**DISTRIBUTION AND HABITAT**. Red-spotted Toads are found from Colorado and Arizona to Mexico, living at elevations ranging from below sea level to nearly 2,000 meters. They thrive in rocky streams in dry areas but can also be found in well-drained, rocky soils away from water. While they are mostly active at night, they can be seen during the day by turning over rocks. They prefer habitats like Sonoran Desert Scrubland and Mojave Desert Scrubland but also inhabit various other environments, including forests.

**DIET.** Their diet mainly consists of invertebrates, with ants being the most frequently consumed prey, followed by termites and beetles. While the size of their prey doesn't significantly relate to the toad's body size, ants and beetles are still the most common food items.

**PREDATORS and DEFENSE**. Red-spotted Toads face threats from snakes, birds, and small mammals. They have a defense mechanism: their skin and parotid glands contain chemicals that can produce hallucinogenic effects, potentially deterring predators.

**FIELD NOTES,** In south-central Arizona, Red-spotted Toads breed alongside other species like Woodhouse's Toads and Lowland Leopard Frogs in the spring, and with Colorado River Toads and Couch's Spadefoots in the summer. Notably, Red-spotted Toad larvae struggle to survive when competing with the larvae of Mexican Spadefoots due to high rates of predation.

---

## Woodhouse's Toad

*Anaxyrus woodhousii* (Girard, 1854) (SE) Figure 14.

**IDENTIFICATION**. Adults can reach 130 mm SUL, with males (83 to 88 mm SUL) smaller than females (93 to 104 mm SUL). Woodhouse's Toad is a typical toad with drab coloration and short legs. It is a medium-sized toad with an olive-brown or green-brown dorsum and a white mid-dorsal stripe. The prominent cranial crests often contact parotoid glands; the white belly has scattered dark pigment, no paired scapular blotches, and a white vertebral stripe. This combination of characters will distinguish this species from other Arizona anurans. It can be distinguished from the Great Plains Toad by the absence of a pair of scapular blotches and from the Arizona Toad—with which it will hybridize—by its elongated parotid glands, well-developed cranial crests, and the pale throat of the male.

**VOCALIZATION.** The call is an explosive, wheezy, nasal snore, w-a-a-a-ah, lasting one to four seconds at a frequency of 1.4–2.0 kHz and rapidly pulsed at 80–200 p/s.

**DISTRIBUTION and HABITAT.** The distribution is from south-central Washington eastward to North Dakota and southward into Durango, Mexico. The elevation range is sea level to 2,600 m ASL. While this species evolved in the Great Plains, it is thought to have expanded its range into Arizona in the mid-to-late Pleistocene with the expansion of grasslands.

Lower-elevation populations occur along more extensive washes, canals, streams, urban ponds, golf courses, and agriculture areas with irrigation. Higher-elevation populations are associated with ponds, lakes, and reservoirs. Thus, they adapt well to human-modified habitats. It occurs in Semidesert Grasslands, Plains and Great Basin Grasslands, and Interior Chaparral.

**REPRODUCTION**. In areas of permanent water, they breed on warm (15°C or above), rainless nights in the spring (February to June), while populations in open desert flats breed

Figure 14. Woodhouses' Toad from the San Pedro Valley. Photography Rene C Clark.

immediately after monsoon storms (June to September). Standing water is preferred for breeding, including pools in river channels, artificial ponds, reservoirs, or cattle tanks in open desert flats.

Eggs are small (3 to 4 mm) and laid in long gelatinous strands. Clutches of up to 28,000 eggs are known for midwestern populations. The larvae are dark brown or grey with mottling and gold flecking; tadpoles transform at about 23 mm. Large aggregations of larvae are often observed resting in standing water on muddy substrate. In central Arizona, the larval period was eight weeks for pond breeding populations (31 March–1 June), whereas development lasted as little as five weeks in Oklahoma.

Males call in shallow water or on the shore at or near the water's edge and defend the calling site. Adults in arid regions may reside near streams or ponds used for breeding year-round. Calling males had a body size range of 69–98 mm, and females had body lengths of 84–109 mm.

**METAMORPHOSIS.** Begins when tadpoles reach a total length of about 30 mm. Recently metamorphosed toadlets average body lengths of 12–15 mm. Tadpoles may form large aggregations in standing water.

**DIET** The diet includes a variety of invertebrates, including isopods, scorpions, centipedes, and spiders; representative insects include orthopterans, lepidopterans, dipterans, hemipterans, beetles, and hymenopterans. They may be best observed at Historic Canoa Ranch in the study area.

**FIELD NOTES.** I have seen this toad only at Historic Canoa Ranch.

## Sonoran Desert Toad

*Incilius alvarius* (Girard in Baird, 1859) (SE) Figure 15-18

**IDENTIFICATION.** Adults may be as large as 190 mm SUL. Males reach 156 mm, and females attain 190 mm SUL. The Sonoran Desert Toad has large parotoid glands, large glands on the legs, a large white wart at the angle of the jaw, and relatively smooth skin for a toad. All these traits combine to distinguish this toad from other Arizona anurans. Newly metamorphosed *I. alvarius* may be mistaken for Red-Spotted Toads. However, they tend to be tan instead of olive green and lack the white swelling or raised area at the corner of the mouth that will eventually become a tubercle.

**LARVAE.** The tadpoles are mottled gold to brass brown and may reach a total length of 56 mm. The body is slightly depressed, the eyes are dorsal, the anal is medial, the oral disk is emarginate, and the tail fin ends near the tail-body junction.

**VOCALIZATION.** The call is a low-pitched whistle that lasts a half-second to one second.

**REPRODUCTION.** In September (late in the reproductive season), the smallest mature male measured 108 mm in body length and contained spermatogonia in Pima County, Arizona. Males mature at 80 mm in body length, and one 87 mm female was mature. Breeding may occur only one night, two or three days after heavy rains. Breeding sites include irrigation ditches, stock tanks, and temporary and permanent ponds. Thus, if permanent water is available, the Sonoran Desert Toad may not depend on rainfall to stimulate reproduction. They use the same breeding sites as Couch's Spadefoot, Mexican Spadefoot, Great Plains Toad, Red-Spotted Toad, and Woodhouse's Toad.

Males have at least two strategies for finding a mate. Advertisement calls are used when a few other males are present, but males may rely more on actively searching for a mate when the competition from other males is great.

Clutches contain up to 8,000 eggs, with the eggs enveloped in tubes of colorless jelly. Eggs average 1.6 mm in diameter and are packed between about 5–7/cm. The time to metamorphosis is less than four weeks.

**DISTRIBUTION.** The species ranges from southwestern New Mexico westward across southern Arizona to southeastern California, southward into northern Baja, and south to Sonora, Mexico. It is absent from the lower Colorado River area. The elevational range is from sea level to 1,600 m ASL. It is primarily in deserts but ranges from mesquite-creosote bush lowlands, arid grasslands, and rocky riparian zones with sycamore and cottonwoods to oak-walnut woodlands in mountain canyons. However, it will also use urbanized habitats and take advantage of storm drains, golf course ponds, and other man-made water features, such as swimming pools. This is a semi-aquatic toad and is more aquatic than most other toads. This toad may sit at the bottom of shallow pools. It occurs in Sonoran Desertscrub, Great Basin Desert Scrub, and Semidesert Grasslands.

**DIET.** Sonoran Desert Toads actively forage for food but also hunt from ambush. Their prey includes beetles, grasshoppers, wasps, centipedes, millipedes, ants, termites, solpugids, spiders, snails, scorpions, other frogs and toads, small lizards, mice, and probably any animal they can overpower.

**MOVEMENT.** A radio-tracked adult in the Tucson Mountains moved more than 400 m daily, while another adult remained in the same burrow under a railroad tie from 26 September to 17 June (about nine months). The toad's body temperature ranged from 11.7 to 29.7 °C. It seems likely that the toad may have been dormant during that period below the surface. Sonoran Desert Toads shelter in rodent burrows for most of the year, and at least some individuals become active before the monsoon.

**PREDATORS and DEFENSE.** The large skin glands produce a toxic cocktail of molecules implicated in the poisoning and death of dogs and humans. When encountering a predator, the toad will tilt its body toward the predator or lower its head and raise its posterior body to present its anatomy and toxins. When disturbed, they will stand on all four limbs and run in a very un-froglike behavior.

A dog owner removed a Sonoran Desert Toad from his dog's mouth within ten seconds. Despite the quick action, within 30 minutes, the dog began to salivate profusely, went into

Figure 15-17 Top. An adult Sonoran Desert Toad. Middle. A juvenile toad. Bottom. A recent metamorph of the Sonoran Desert Toad. Photography JCM

Figure 18. An adult Sonoran Desert Toad attempting to stay cool during the heat of a 115-degree day. This massive toad was using an air-conditioning humidifier to stay cool. Note the falling drop of water above the toad. Photography JCM.

convulsions, and died from apparent respiratory failure. In 1986, a five-year-old boy with profusal salivation and continuous seizures was admitted to the University of Arizona Medical Center. Seizure activity started within 15 minutes of his licking a toad. The child survived, but it took a week to return to normal. The Sonoran Desert Toad produces large quantities of the potent hallucinogen S-methoxyN, N-dimethyltryptamine (5MeO-DMT), a molecule that can cross the blood-brain barrier. However, the skin also contains toxic bufotenines (indolealkylamines and their metabolites.) Bufotenine can be present in 3 mg per gram of dry skin. Other research has found new compounds and enzymes involved in the chemical defenses of this toad.

At least two predators, ravens, and raccoons, have learned to remove the toads' highly toxic skin before eating them. Trap cameras set up at the Northern Jaguar Reserve in Sonora captured images of bobcats, coyotes, and badgers interacting with Sonoran Desert Toads. The Bobcat and the Coyote ignored the toads. The Badger was walking off with the toad in its mouth. The invasive Northern Crayfish, *Orconectes virilis*, may also be a significant Sonoran Desert Toad tadpole predator in some localities.

**FIELD NOTES.** A relatively common species. Observations suggest activity may start well before the onset of the monsoon; adults may be active in late April in urban areas. In 2025 I found a road killed specimen on 19 April. This may be triggered by landscape watering or the rare brief shower. Individuals have been observed feeding at night in semi-desert grasslands in June. My (JCM) neighbors have one of these toads that has lived in their yard for several years. During monsoon evenings, it will move about a hundred meters from their yard to my driveway to sit under lights and feed on insects. It returns to my neighbor's yard when the lights are turned off.

# Fleshbelly Frogs

The family Craugastoridae is a clade restricted to the Western Hemisphere. It contains 136 species that range from the southern United States into Central and South America. Craugastorid frogs are ecologically diverse, using habitats that range from deserts to tropical forests. Streicher (2012) found that the Barking Frog, *Craugastor augusti* has minor mitochondrial or nuclear sequence divergence. He suggests the Barking Frog only recently expanded into its present distribution, which may be related to the use of underground refugia during cold, dry periods. Ryan et al. (2015) proposed that habitat fragmentation and the species' small range increase the risk of extinction. The Barking Frog is a habitat specialist restricted to unique and isolated landscape features, limestone escarpments and creosote flats with soft, gypsum soils.

---

## Western Barking Frog

*Craugastor augusti cactorum* (Taylor, 1938) (IM) Figure 19.

**IDENTIFICATION.** Adult males in Arizona are 71–74 mm, and females are 77– 83 mm SUL. The Huachuca Mountain population has adults 47–94 mm, with larger females. Adult coloration is primarily brown. The dorsal skin is granular, dorsolateral folds are present, and a distinctive transverse skin fold is on the back of the head. Toes lack webbing. Large brown spots with pale green centers may be present on the dorsum and sides of the body. Indistinct bars may be present on the limbs. The combined dorsolateral folds and the transverse fold of skin behind the head will distinguish this frog from all other Arizona species.

Males have dark tympanums and dark throats during the reproductive season; females have pink tympanums and white throats year-round. Juveniles are very different in appearance. The head and anterior body are dark brown or black with light speckling. A white transverse band across the mid-body and forelimbs makes juveniles distinctive.

**VOCALIZATIONS**. Often in cavities such as rock crevices but will also call from open situations. The call has been described as a dog's bark, the caw of a raven, and a *whurr*. In Arizona, the call lasts 573 ms. Males often call in sequence, with one individual leading the chorus. It is the only Arizona frog that uses direct development, i.e., there is no tadpole stage. The eggs develop directly into froglets.

**REPRODUCTION.** Eggs are deposited in crevices and fissures or under rocks on moist soils. One of the parents guards the eggs. Clutches contain 6 to 80 eggs; incubation requires 25 to 30 days. The parent's role in tending the eggs has been described as protecting them from predators and desiccation. Observers have described the protecting parent urinating on eggs, presumably to keep them moist. However, this may also pass chemical protection to the eggs to prevent bacterial and fungal infections. It is unclear which parent stays with the eggs. Males have been most often suggested as the guardian; however, this has been questioned because they are highly mobile.

Figure 19 Western Barking Frog. Photography Mike Rochford

**HABITAT and DISTRIBUTION**. They inhabit limestone outcrops of the Madrean Sky Islands of the Sonoran Desert, Dry Tropical Forests, and Tropical Pine-oak forests. They have been documented in the Santa Rita, Pajarito, Huachuca, and Quinlan ranges in Arizona, at 1,280–1,890 m ASL

**ACTIVITY.** Surface activity occurs during the monsoon, most often after a thunderstorm. They are rarely encountered, partly due to their limited and disjunct range but also because they tend to call only during the first few heavy monsoon rains over two to four weeks.

Males may move up to 50 m from their refugia to their calling station. Calling is associated with no wind, high humidity, and low temperature. Males have also been observed calling in the late afternoon. They are usually nocturnal and located when people triangulate on their call.

**DIET.** The diet includes a variety of invertebrates, including camel crickets, field crickets, Gladstone Grasshoppers, long-horned katydids, short-horned grasshoppers, land snails, silverfish, centipedes, scorpions, kissing bugs, spiders, and ant lions.

**DEFENSE BEHAVIOR.** Barking frogs' defense mechanisms include inflating their body and increasing their size to avoid predation. Individuals surviving 11 years in captivity appear to be relatively long-lived.

**CONSERVATION.** Barking frogs have no special protection at the state or federal level. They are considered a Species of Special Concern in Arizona (Arizona Game and Fish Department, 1996a).

**FIELD NOTES.** Restricted to a very specific rock crevice microhabitat. Rarely seen.

# Narrowmouths

Narrowmouths are often referred to as "toads." However they are all in the family Microhylidae. They are primarily tropical in South America, Africa, Madagascar, Southeast Asia, Indonesia, New Guinea, and northeast Australia, but some extend into warm temperate North America and Africa. About sixty genera contain more than 690 species.

They range from some of the smallest known anurans (10 mm) to medium-sized anurans (100 mm). The body form is frequently tear-drop-shaped, with a narrow, pointed snout and somewhat rounded body. However, some are more treefrog-like, with expanded digital tips. The two genera in the United States are *Gastrophryne* and *Hypopachus*. The family last shared an ancestor with the ranid frogs 116 million years ago.

North American narrowmouths of the genus *Gastrophryne* have been long viewed as two species, with an additional three species in southern Mexico and Central America. Edward Taylor described a third Mexican species, *Gastrophryne mazatlanensis,* in 1943. Three years later, Hetch and Matalas synonymized *G. mazatlanensis* with *G. olivacea* because they found individuals they considered intermediate. Blair (1955) examined the calls of both species across their combined distributions, found minimal hybridization in the overlap zone, and discovered most individuals in the overlap zone were one species or the other.

Streicher et al. (2012) examined mitochondrial variation with respect to all the currently recognized species of the genus *Gastrophryne*. They found the Great Plains Narrowmouth and Taylor's Sinaloan Narrowmouth to be distinct species, differing by 2.2% genetically.

---

## Sinaloan Narrowmouth

*Gastrophryne mazatlanensis* (Taylor 1943) (AS) Figure 20.

**IDENTIFICATION.** Adults range from 19 to 36 mm SUL. Males are smaller than females. New metamorphs are 9 to 10 mm. A small, stout frog with a pointed snout and a transverse skin fold behind the head. Its limbs are relatively short. The tympanum is not visible. The skin is smooth and dark brown to brown-green. The dorsum has small, scattered black and cream spots with larger, irregular black pigmentation areas on the hind legs. There is an enlarged metatarsal tubercle on the foot. Males have a black gular region and tubercles on the chin. A black bar extends across the femur and tibia when the limb is folded. The head and body shape, small size, and transverse skin fold combine to make this frog readily identifiable.

**VOCALIZATION.** The call is a high-pitched buzzing that may be mistaken for an insect call or the call of the Sonoran Green Toad. Males call from extremely small crevices near water, under rock over-hangs while in the water, and from submerged leaf litter, making them exceptionally difficult to locate. Males will sometimes call before sunset. Reports suggest skin toxins of the Sinaloan Narrowmouth will kill other frogs housed in the same container.

**REPRODUCTION.** The Sinaloan Narrowmouth is an explosive monsoon breeder. Calling males are often concealed in vegetation; however, they may call from floating vegetation or the edge of the land adjacent to the water. Breeding sites are frequently dense stands of floodplain mesquite and include small puddles such as tire tracks or cattle hoof prints. Eggs can be seen as rafts of eggs floating on the surface. Two clutches contained 175 and another 200 eggs. The gelatinous envelope was 4 mm, and the ovum was 1.2 to 1.4 mm. Tadpoles transform in the first season.

**LARVAE.** The tadpoles are tiny, 31 mm in total length. The body is flat, compressed and elliptical, and the tail tip is acute. They are black with a ventral midline stripe and a ventrolateral stripe on each side that converge anteriorly. The tail has a white stripe that breaks into spots close to the body. Spots are most abundant on the dorsal fin. The tadpoles will stop swimming and float on or near the surface.

**HABITAT.** The Sinaloan Narrowmouth Toad's habitat is as unique as its physical characteristics and vocalizations. Pima and Santa Cruz Counties, it is in semi-desert grassland

Figure 20. The Sinaloan Narrowmouth. Photography JCM

and occurs in Madrean Woodland in the Pajarito and Patagonia Mountains. It is present in the Santa Ritas, the Patagonia Mts, and the surrounding habitats. The habitat ranges from low-elevation creosote flats through grasslands to oak-woodland communities near Ruby, Arizona, at an elevation of more than 1200 m on the eastern margin of the study area (San Xavier and the vicinity of Pena Blanca Lake). It has been observed in southwestern Santa Cruz County, just east of the Buenos Aires Refuge boundary.

**DISTRIBUTION.** It is distributed from south-central Arizona to the west coast of Mexico as far south as Nayarit. In Arizona, it is known from the Vekol Valley (southern Maricopa County) southward to the Mexican border.

**FIELD NOTES.** This tiny frog may travel long distances – more than two kilometers overland from breeding sites. This species was observed in my yard, a development in semi-desert grassland within a few minutes after a particularly heavy rain in Sahuarita, Arizona (in 2017). Suggesting that it is more common and widespread than previously realized.

# Treefrogs

Treefrogs, chorus frogs, and cricket frogs form the family Hylidae, a cosmopolitan family with its most extraordinary diversity in the Western Hemisphere. While they are called treefrogs, and many arboreal lineages occur within the family, some have diversified to become burrowers, others are aquatic, and some are entirely terrestrial. Hylidae is one of the most prominent anuran families, with almost 1,000 species recognized at present, ranging from small (12 mm) to large (145 mm) frogs. Most have adhesive toe discs that contain a piece of cartilage (the intercalary cartilage), offsetting the terminal phalanx. This bent-toed structure may aid in climbing, but it has been lost in some lineages. Arizona has at least six species of hylids in four genera. Two of these occur in the focus area.

---

## Canyon Treefrog

*Dryophytes arenicolor* (Cope, 1866) (SE) Figure 21.

**IDENTIFICATION.** Adult frogs range from 32-57 mm in SUL. A treefrog with rough, warty skin and webbing between toes is well developed but does not extend to the hind leg's fifth toe, and toe pads are enlarged. Its dark-edged light spot beneath the eye distinguishes it from other Arizona hylids. Dorsal color pale white-grey to brown, to grey olive, it may be almost black when cold; scattered spots and blotches of darker pigment present; a transverse bar or broad triangle sometimes present on the head; legs with dark bars; ventral surface whitish or cream, with yellow or orange on the groin, femur, tibia, and axilla; iris with metallic yellow or cream. Albino specimens have been found, and albino frogs have red pupils with white irises. The following description of *arenicolor* is based on specimens collected in southeast Arizona. Tadpoles are black and become olive green with a yellow belly with age, attaining 43 mm total length. The tail may have transverse bands or marbling—Iris olive with gold markings.

**VOCALIZATION.** Advertisement calls are hollow, nasal, and explosive, lasting only 1-3 seconds. The call has been described as a sheep-like *ba a a*, a quacking duck, and an *r,r,r,r.* Females answer the male's call with a lower note. The rocky habitat may enhance the calls from boulder-strewn streams.

**DISTRIBUTION.** Specifically found in parts of southern and western Colorado. Isolated

Figure 21. A male Canyon Treefrog calling in a stream. JCM

populations are found in Southeastern Arizona, northeastern New Mexico, and certain regions of Texas.

**HABITAT**. It is associated with streams and streambeds transecting arid environments. Often in canyon bottoms. It generally lives on the ground but can be found in trees and clinging to boulders. They are usually seen sunning on boulders in or near the stream during the day. Canyon Treefrogs are not commonly found in standing water but prefer more humid environments, such as crevices in rocks and under logs.

It occurs in Arizona Upland Desert Scrub, Semidesert Grassland, Interior Chaparral, Madrean Evergreen Woodland, Great Basin Conifer Woodland, and Petran Montane Coniferous Forests.

**DIET.** This species takes a variety of insects, including beetles, ants, and caterpillars. Termites and beetles were the most important prey, but volumetrically, beetles and orthopterans were most important.

**REPRODUCTION**. The breeding period ranges from March to July but can be extended due to insufficient rainfall. Most breeding activity is likely to occur in spring or early summer. Although a large chorus was encountered on July 1, followed by 13 mm of rain that day. On July 17, small tadpoles were found. Most breeding that year took place during this period, for there was little calling later, and no females were found after July 8. Eggs are 3.8-5.0 mm in diameter and attached to debris and rocks at the bottom of rocky pools. Eggs may be single or in clumps. Tadpoles transform in 40 to 75 days of age and about 15-18 mm in body length. Transformation occurs from early June to mid-August. Newly transformed froglets are about 17 mm.

Canyon Treefrogs are relative easy to observe. Their abundance along boulder-strewn streams and the habit of sunning themselves during the day provides opportunities to see them. Their mottled pattern gives them camouflage, but they look like a small bump on an otherwise smooth boulder. Choruses are often small near isolated pools along streams.

---

## Arizona Treefrog

*Dryophytes wrightorum* (Taylor, 1938) (VU) Figures 22-23.

**IDENTIFICATION.** Adults range from 37 to 48 mm SUL; metamorphs are 10 to 13 mm SUL. Adults are green, with a dark brown or black stripe extending from the snout, through the eye, and onto the body. The ventral side of the groin and upper leg are orange to gold with a greenish tint. The upper lip has a white stripe but not the dark-edged white spot seen in the Canyon Treefrog. The skin is smooth. The new metamorphs are brown but change to green in a few days. The fingers lack webbing, but the toes are webbed.

The call is a low-pitched metallic trill that lasts 0.15 to 0.24 seconds and is repeated. Choruses are usually short.

**DISTRIBUTION** The species occurs from central Arizona and western New Mexico southward in the Sierra Madre Occidental to Michoacán, Mexico. It appears on the central plateau north of the Mogollon Rim in Arizona, with disjunct populations in the Huachuca Mountains and Canelo Hills in southeastern Arizona. There is a record from the grasslands of northern Sonora, Mexico's San Pedro River drainage, just south of the Arizona border. There is also a record from the oak-grassland bajada at about 1410 m on the east side of the Huachuca Mountains in southeastern Arizona, USA, suggesting that the Huachuca population reaches substantially lower and drier elevations than the Mogollon Rim population.

**HABITAT.** Outside the breeding season, they are in pine-oak and coniferous forests and meadows near streams between 900 and 2,900 m ASL. During the breeding season, they move to water, but their migrations are poorly documented. They overwinter in forest floor debris.

**REPRODUCTION.** Monsoon rains trigger males to call a few nights after the rains start. Breeding sites tend to be grassy ponds but may also use permanent water or wherever rainwater collects. Larvae have been reported from 21 June to 6 November, and metamorphs are present in September and October. They share breeding sites with the Arizona Toad, Canyon Treefrog, Boreal Chorus Frog, leopard frogs, and several species of spadefoots.

Adult males are 24–44 mm SVL (Wright and Wright 1949). The mean body size of females was significantly larger than that of males. Five adult females (mean SVL = 40.0 mm and a range of 36–44 mm.

**LARVAE.** At hatching, the larvae are 4.9 to 5.2 mm. in total length. The maximum total length attained by any tadpole was 38 mm., and the maximum body length was 12.7 mm. The newly hatched tadpole is pale yellow in the abdominal region, where the yolk shows through, and yellow-brown on the rest of the body and tail. A characteristic feature of *wrightorum* is the early development of melanophores both on the body and in the tail fin. The largest tadpoles have at least a stippling of dark brown pigment in the anterior-most parts of the tail fins. This is always a green frog, whereas the related species *D. eximia* and *D. regilla* have both green and brown phases and mixtures of the two colors.

**DIET.** A variety of invertebrates, beetles, spiders, and small earthworms are taken as prey. Tiger salamander larvae use some of the same habitat as these frogs and eat the tapoles. Other predators include water bugs, crayfish, bullfrogs, and gartersnakes. Skin secretions have been reported to cause a burning sensation on mucus membranes.

**GEOGRAPHY.** *Dryophytes wrightorum* is primarily found in the southeastern regions of Arizona, particularly in areas with perennial water sources such as streams and ponds. Phylogenetic studies have shown that this species is closely related to *Dryophytes arenicolor*, with evidence suggesting that they share a common ancestor. The genetic differentiation between these two species is influenced by their respective habitats and ecological niches. For instance, *D. wrightorum* tends to inhabit higher elevation areas with specific moisture requirements critical for its reproductive success. This specialization may have arisen due to historical climatic changes that created isolated populations, allowing for adaptive divergence.

Figure 22. Arizona Treefrog. Photography Bill Lipinsky.

Figure 23. A juvenile Arizona Treefrog. Photography Stephanie M. Cobbold.

On the other hand, *Dryophytes arenicolor* is more commonly associated with canyon habitats and is adapted to a broader range of environmental conditions, including varying moisture levels. This species has been shown to exhibit introgressive hybridization with *D. wrightorum*, which complicates the understanding of their evolutionary trajectories. The hybridization events between these two species suggest that they have overlapping ecological requirements, which may facilitate gene flow and influence population dynamics in shared habitats.

Moreover, the population structure of *D. wrightorum* has been characterized by isolation-by-distance patterns, indicating that geographic separation plays a significant role in genetic differentiation. This isolation is particularly evident in fragmented habitats where human activities have altered the landscape, leading to declines in suitable breeding sites for both species. The conservation of these treefrog populations is thus critical, as habitat loss and fragmentation threaten their survival and genetic diversity.

In summary, the origins of *D. wrightorum* and *D. arenicolor* in Arizona are shaped by their evolutionary history, ecological adaptations, and interactions through hybridization. Understanding these dynamics is essential for effective conservation strategies aimed at preserving these unique amphibian populations in the face of ongoing environmental changes.

## True Frogs

Ranid frogs are sometimes called true frogs because Carlus Linnaeus applied the Latin name *Rana*, to virtually all the frogs he described in the mid-18th century. Ranids have the greatest distribution of any frog family and are represented in North America, South America, Europe, Asia, the Indonesian Archipelago, Madagascar, Africa, and Australasia. The family has more than 1000 species ranging from some of the smallest anurans (10 mm) to the largest frog, *Conraua goliath* (adult size up to 320 mm). The major ranid clades show a close association with the Gondwanan plates, suggesting that the breakup of Gondwana played a significant role in the current distribution of ranid frogs. Ranid frogs are the sister to Microhylidae, as the two families last shared an ancestor about 111.9 MYA.

The populations of ranid frogs, particularly in Arizona have steadily declined for the past fifty years. This decline is not only due to habitat loss and predation by the invasive American Bullfrog but also to the deadly amphibian chytrid fungus, *Batrachochytrium dendrobatidis* (*Bd*), the ranavirus, red-leg syndrome, pesticides, endocrine disrupters, and a receding water table. This trend calls for immediate attention and action to prevent further decline.

*Lithobates* is a controversial genus of true frogs within the family Ranidae. Initially defined as a subgenus, it now comprises most North, Central, and South American frogs previously assigned to the genus *Rana*. The American Bullfrog (and most other North American ranids) are assigned to this genus.

Alternatively other authors have retained the genus *Rana* for virtually all North American species In summary, the controversy over the classification of *Rana* versus *Lithobates* reflects broader issues in taxonomy, including the balance between scientific accuracy and the need for stability in nomenclature. The resolution of this debate will require ongoing dialogue and collaboration among researchers to ensure that naming conventions accurately reflect evolutionary relationships while remaining accessible to the broader community

---

## American Bullfrog

*Lithobates catesbiana* (i) (SE) Figures 25.

**IDENTIFICATION**. Adults reach at least 200 mm in body length (usually 90-150 mm) and can reach weights of 800 g. The mean SUL length for males is 152 mm (range 111-178), and for females, the mean SUL is 162 mm (range 120-183 mm). It can be recognized by the absence of dorsolateral folds and an exceptionally large tympanum (larger than the eye in males). The tips of the digits are blunt, and the webbing between the toes is well-developed. The dorsal skin is rough with scattered minute tubercles. There is a prominent supratympanic fold. The males have pigmented nuptial pads. The vocal sacs open on each side of the ure e, from brown (when cold) to shades of green (when warm). The dorsum is green and mouth, the rictus. Coloration varies widely depending on the frog's locality and temperature. Skelly et al. (1999) found bullfrogs restricted to open canopy ponds with permanent water. Bullfrogs prefer relatively high body temperatures (26–33$^{0}$C) and thermoregulate, basking and altering their posturing to control body temperatures. Adults remain inactive until water temperatures approach 15$^{0}$C. Unlike many other frogs, bullfrogs can coexist with predatory fishes due to their skin sections, which make them distasteful or toxic to fish

**HABITAT**. This is a highly aquatic frog that can make long trips overland to find habit-usually permanent stagnant or slow moving water choked with vegetation.

**DIET.** Bullfrog feeding behavior has been described using voracious, opportunistic, and cannibalistic adjectives. They are usually ambush predators and attack any live animal smaller than themselves (Bury and Whelan 1984). Bullfrogs may locate and eat smaller frogs by following the sounds of chorusing frogs or distress calls from other frogs. Bullfrogs will also eat carrion. Small individuals eat insects, while larger individuals eat frogs, crayfish, and mice. In Arizona, most stomach contents of adults consisted of invertebrates such as snails

Figure 24. Bullfrogs are invasive in Arizona and a threat to the survival of other amphibians and reptiles. This one is two colors because it is warming up. The colderpart of the body is brown the warmer part is green. Photography JCM

and insects. Bullfrogs also prey on small snakes, including the threatened Mexican Garter Snakes. The eradication of these alien frogs is essential to maintaining native biodiversity.

Western North America and many countries in Europe, Asia, and South America.

The species ranges from southeastern Arizona and southwestern New Mexico into Mexico along the eastern slopes of the Sierra Madre Occidental. The upper surface of the body is greenish olive to dark olive, with many minute specks of black; the sides of the tail have spots (up to 2 mm. in diameter) of yellow; the ventral surface of the body yellow to white, sometimes mottled with dusky, never iridescent; iris golden yellow.

**REPRODUCTION**. Sexual maturity varies with latitude and altitude. It takes 1–2 years for males and 2–3 for females to mature. They may live up to 10 years in the wild; captives have been reported to have lived for 16 years. Eggs are laid in thin rafts on the water surface. Females may produce clutches of 48,000 eggs or more that hatch in 3–5 days. Tadpoles favor warm-water environments, and development time is temperature-dependent. Development can be as short as a few months in the south to three years at higher latitudes and elevations.Reproduction may change from year to year depending on the temperature; a cold or warm spring-summer may postpone or advance breeding dates. Older females may produce two clutches per year, with second clutches containing substantially smaller eggs (Howard 1978b).

**PREDATION.** Breeding occurs in late spring and early summer at southern latitudes and later in the summer at northern latitudes. This is also true for other locations, and the timing of reproduction. Bury and Whelan (1984) noted Bullfrogs eggs need higher temperatures (15-32C)to develop than other North American frogs. In Nebraska, largemouth Bass and Green Sunfish would only maintenance feed on Bullfrog tadpoles when no other prey was available. Black Bullheads choose to starve rather than feed on Bullfrog tadpoles. The ability of Bullfrog tadpoles to co-exist with predatory fish may result from the tadpoles being unpalatable.

## Chiricahua Leopard Frog

***Lithobates chiricahuensis*** (Platz & Mecham, 1979) Figure 25.

Other names for this frog include *Rana chiricahuensis*.

**Identification.** Adults range from 51 to 137 mm, and tadpoles metamorphose at 30 to 40 mm SVL. Females are larger than males. They have an olive or brown dorsum with small black spots without a pale border, an incomplete or absent white lip stripe, and an interrupted dorsolateral fold. The skin is rugose with dorsal tubercles, and the throat is a mottled grey. Because this is a species complex, the morphology and the call can be expected to vary across the range. The call is a single note that lasts 1 to 2 seconds and sounds like a snoring trill that is made while submerged, or a semi-croak, or a very low croak.

**DISTRIBUTION.** The species ranges from southeastern Arizona and southwestern New Mexico into Mexico along the eastern slopes of the Sierra Madre Occidental. In Arizona, it is restricted to the southeast corner of the state.

**HABITAT.** Includes semi-desert grassland, Madrean woodland, pinyon-juniper coniferous forest, and montane coniferous forest. It is usually associated with streams flowing through these habitats, as well as ponds. Overall, this is a highly aquatic species that does not wander far from water.

**REPRODUCTION.** This synchronous with the monsoon but may occur at any time during the warmer months (February to September). Populations at higher elevations (above 1,800 m) will reproduce later in the season than those at lower elevations. Eggs are attached to vegetation. Eggs hatching late in the season will have their tadpoles overwinter.

Field workers in the 1960s characterized Chiricahua Leopard Frogs as abundant in well-watered sites in southeastern Arizona. However, most populations have declined precipitously, while they remain locally abundant at some sites. The first report of declining Arizona leopard frogs was published in the mid-1980s. The problem seemed to be due to introduced fish and the introduced bullfrog, which preyed upon the native leopard frogs. Also, the habitat was degraded at many locations.

The reintroduction of the Chiricahua Leopard Frog to portions of its historic range started in 2003 and was combined with an active program to remove bullfrogs from some localities. The risk of bullfrog extinction to the re-established populations was considered low because the species has excellent dispersal abilities (Chandler et al., 2015). Today, the leopard frog populations have improved and will likely continue to thrive, unless or until, the bullfrog returns. Howell et al. (2016) examined the survival rate of this frog at a reintroduction site and found it low (27%) compared to other reintroduced ranid frog populations. They attributed their findings to emigration of the frogs away from the study site.

Multiple species appear to be covered under this name at present. The Ramsey Canyon population has been described as *Lithobates subaquavocalis*. The Las Vegas Leopard Frog was originally known from the Las Vegas Valley in southern Nevada, where it has been extirpated. The populations of leopard frogs from the western Mogollon Rim previously recognized as the Chiricahua Leopard Frog are morphologically distinct from the populations of Chiricahua Leopard Frogs in southern Arizona, New Mexico, and Mexico. Hillis et al. (2005) suggest the Mogollon Rim populations may be conspecific with the Vegas Leopard Frog, *Lithobates fisheri*. Prior to that it was considered extinct until Hekkala et al. (2011) provided evidence that the morphologically distinct Chiricahua Leopard Frog from the western Mogollon Rim was actually the Las Vegas Leopard Frog. They obtained

Figure 25. A Chiricahua Leopard Frog. JCM

genetic information from a California Academy of Science specimen preserved in 1913 and determined that it is a member of one of the two lineages, which is extant along the western Mogollon Rim and White Mountains of central and eastern Arizona and extreme west-central New Mexico. The Las Vegas Leopard Frog may have emerged from extinction and expanded distribution with this new knowledge. Further investigation, however, may show it is a distinct form.

The type locality for *Lithobates chiricahuensis* is Herb Martyr Lake (elev. 1,768 m), 6 km west of Portal, Coronado National Forest, Cochise County, Arizona. Thus, the frog that eventually ends up with the name *Lithobates chiricahuensis* will be from the southeast corner of the state.

The Chiricahua Leopard Frog is a closed-season species. Collection of leopard frogs requires a scientific permit or a similar permit (Arizona Game and Fish Department, 2001). The species is listed as threatened under the Endangered Species Act of 1973.

**CONSERVATION.** The first report of declining Arizona leopard frogs was published in the mid-1980s. The problem was attributed to introduced fish and the introduced bullfrog, which preyed upon the native leopard frogs. Also, the habitat was degraded at many locations.

The Vegas Valley Leopard Frog, *Rana fisheri* was described by Stejneger 1893 7:227 based on a holotype from Vegas Valley, Clark County, Nevada. The suggestion that the extinct Vegas Valley leopard frog (also known as *Lithobates fisheri*) may have been synonymous with one of several declining species complicated recovery planning for imperiled leopard frogs in the southwestern United States. To address this concern, Hekkala et al. (2011) and later explained by Dodd (2013) found the Las Vegas Leopard frog still exists on the Mogollon Rim, and was confused with the Chiricahua Leopard Frog for those Mollogon Rim populations. The populations of southeastern Arizona were the Chiricahua Leopard Frog (*Lithobates chiricahuensis* Platz and Mecham 1979). Thus the populations in the focus area are correctly called the Chiricahua Leopard Frog, *Lithobates chiricahuensis.*

## Tarahumara Frog

*Lithobates tarahumarae* (Boulenger, 1917) (IM) Figure 26.

**IDENTIFICATION.** Adult males reach 88 mm, adult females to 113 mm. Males mature at about 64 mm, females at about 67 mm. Newly metamorphosed frogs are 35–40 mm. Adults are dark brown with small black spots on the dorsum. The skin can be smooth or rugose, and the belly is white, with some yellow in the groin area. A tympanic fold extends from the corner of the eye to the insertion of the front leg. The tympanum is smaller than the eye. Most specimens lack dorsolateral folds, but some specimens show a trace of these folds. The fingers and toes end in small, round disks.

**VOCALIZATION.** Long considered voiceless, at least four different calls have now been recorded. They have been described as a snore, a soft whine, an *eep* or *eep-up*, and a croak or squawk. Visually, it is perhaps most likely confused with the invasive American Bullfrog. Still, bullfrogs usually lack distinct bars on the hind legs, have a conspicuous tympanum, and no trace of dorsolateral folds.

**HABITAT.** The Tarahumara Frog lives in canyons with rocky streams and plunge pools at the base of waterfalls. It is associated with tropical deciduous forests, foothills, thornscrub, semi-desert grassland, oak woodland, and pine-oak woodlands. It ranges from extreme southern Arizona southward to northern Sinaloa and southwestern Chihuahua, Mexico.

The frogs emerge from winter torpor in early April at temperatures above 10°C. In Arizona, they can be active day and night from April through October. During the dry season (April-May and into June), they congregate in areas of perennial flow, drought-resistant plunge pools, and tinajas. The frogs expand their activity when rainwater increases stream flow in June or July. Perennial plunge pools in streams with low mean flow and relatively steep gradients provide the best breeding sites.

**REPRODUCTION** occurs towards the end of the dry season and in late summer. Spherical egg masses containing about 2,000 eggs are deposited in pools, where they sit on the bottom and become covered with sediment. The tadpoles may take up to two years to metamorphose. However, metamorphosis may occur in three months in captivity with a constant food supply and optimal temperatures. Males lack vocal sacs, but both sexes and juveniles make various sounds when water exceeds 15°C.

Figure 26. The Tarahumara Frog. National Park Collection.

**MOVEMENT.** The Tarahumara Frog is capable of long-distance dispersals overland and along drainages. One female was marked and released in 2004 as part of a re-establishment program and moved at least 1.6 km. This same frog moved about 3.4 km across rugged terrain a second time.

Berry Campbell (1931) described finding the Tarahumara Frog for the first time in the USA. "On June 18, 1931, near Pena Blanca Springs, Santa Cruz County, Arizona, Campbell was fortunate in finding a colony of the Mexican frog, *Rana tarahumarae* Boulenger. Dr. Remington Kellogg kindly examined a specimen forwarded to the National Museum, who pronounced it that species."

**CONSERVATION.** By the mid-1980s, fifty years after its initial discovery in Arizona, the Tarahumara frog was extirpated from the state. The last observation of Tarahumara Frogs in Arizona was in May 1983 in Big Casa Blanca Canyon in the Santa Rita Mountains. The reasons for the frog's disappearance are unclear, but heavy metal poisoning, disease, and habitat loss were suspected. It was also suggested that it may have been the first frog extirpated from the United States due to *Bd (Batrachochytrium dendrobatidis).* Efforts to re-establish the frog at some Arizona and Sonora locations started early in the 21st century.

Arizona Game and Fish Commission Order 41 prohibits the collection of Tarahumara Frogs from the wild in Arizona. Arizona State law makes it illegal to harass, harm, pursue, hunt, shoot, wound, kill, trap, capture, or collect this species

---

## Lowland Leopard Frog

*Lithobates yavapaiensis* (Platz and Frost, 1984) (IM) Figure 27.

**IDENTIFICATION.** Adult males are 50 to 57 mm, and adult females are 59 to 72 mm. The maximum size reported is 87 mm. Newly transformed frogs are 25 to 29 mm. The dorsum is brown to olive-brown with dark brown or pale brown spots that lack a pale edge; the head lacks spots; the skin is smooth; the pale-colored dorsolateral folds are weak, interrupted, and directed medially; and the posterior surface of the thighs have a dark reticulated pattern; the upper thighs are barred. This combination of traits distinguishes this frog from other Arizona leopard frogs. Vocalization. Males attract females by emitting a call consisting of a series of low pulses lasting 3 to 8 seconds.

**DISTRIBUTION.** The species ranged from southwestern New Mexico westward across southern Arizona to Mohave County, Arizona, and was present in the western Grand Canyon. It was also known from southern California and Sonora, Mexico. However, its distribution is significantly reduced and has been extirpated from a significant portion of its former range. It has not been seen in California since 1965, and it has not been seen in New Mexico since1985.

**HABITAT.** Rivers, permanent and intermittent streams, beaver ponds, springs, cattle tanks, livestock drinkers, canals, irrigation sloughs, wells, abandoned swimming pools are all used by the Lowland Leopard Frog. Six localities in Arizona ranged in elevation from 658 to 1,134 m and had average annual rainfall of 233 to 480 mm per year. Fragmentation and human manipulation of the water table and drainage basins to water crops and cattle are primarily responsible for its reduced distribution.

**REPRODUCTION.** Egg masses have been seen between January and late April and again in October. Females deposit clusters of eggs attached to submerged vegetation or a substrate of bedrock or gravel, often near the water's surface. Eggs reportedly hatch in 15 to 18 days, and tadpoles may metamorphose in the same year they were laid or overwinter. Thus, the time to metamorphosis may be as short as three or four months or as long as nine months.

**CONSERVATION.** Lowland Leopard Frog populations are vulnerable to widespread mor-

Figure 27. Lowland Leopard Frog. Photography JCM

tality and extirpation due to flooding, invasive species, and habitat loss from draining and siltation of breeding pools. There is no open season for this species.

# Reptiles

Testudines (turtles), Crocodylia (crocodilians), Squamata (lizards and snakes), and Rhynchocephalia (the Tuatara) are the four orders that makeup living reptiles. Approximately 12,000 extant reptile species are included in the Reptile Database as of May 2023. Reptiles are tetrapod vertebrates, animals that either have four limbs or are descended from animals with four limbs. Reptiles do not have an aquatic larval stage, in contrast to amphibians. All turtles and crocodilians lay eggs; most lizards and snakes are oviparous, but some are viviparous. In these instances, the embryo grows inside the mother using a placenta-like structure like a mammal instead of being enclosed in an eggshell. Reptile embryos are protected and nourished by membranes, which allows them to reproduce on land.

## Turtles & Tortoises

Turtles are characterized by a unique shell developed mainly from their ribs. There are 360 living and recently extinct species of turtles, including land-dwelling tortoises and freshwater terrapins. They are found on most continents, some islands, and most of the oceans, as in the case of sea turtles. Like other amniotes (reptiles, birds, and mammals), they breathe air and do not lay eggs underwater, although many species live in or around water. Turtle shells comprise a domed carapace; the underside is the flatter plastron or belly plate. Its outer surface is covered in scales made of keratin, the same material composing hair, horns, and claws. The carapace bones develop from ribs that grow sideways and develop into broad flat plates that join up to cover the body. Turtles are ectotherms or "cold-blooded," meaning that their internal temperature varies with their direct environment. They are generally opportunistic omnivores and mainly feed on plants and animals with limited movements. Many turtles migrate short distances seasonally. Sea turtles are the only reptiles that migrate long distances to lay their eggs on a natal beach. Turtle habitats around the world are being destroyed. As a result of these pressures, many species are extinct or threatened with extinction.

# Pond Turtles

The family Emydidae, a widespread lineage in the Western Hemisphere and Eurasia, are called pond turtles or freshwater turtles. Still, some are terrestrial, rarely entering the water. With 122 species in 28 genera, emydids can be aquatic, semi-aquatic, or terrestrial and most are omnivorous. Two species are native to Arizona: the aquatic Western Painted Turtle and the terrestrial Ornate Box Turtle. While these two species are in the same family, ecologically, they are quite different and shared an ancestor at about 41 MYA. An other member of the lineage has been introduced to Arizona: the Slider, *Trachemys scripta.* It is unknown from the area covered in this volume, but given the large number of people who move these turtles around, it's only a matter of time until they arrive, or they may already be here.

The genus *Terrapene* comprises North American box turtles. In appearance and terrestrial habits, box turtles superficially resemble tortoises. The genus includes about 12 species and subspecies. They all possess a high domed carapace and a hinged plastron, allowing them to retract their heads and legs and tightly close their shells for protection against predators.

---

## Ornate Box Turtle

*Terrapene ornata* Agassiz, 1857 (IM) Figure 28.

**IDENTIFICATION.** Adult females grow to 154 mm, adult males to 130 mm. Hatchlings are 28–32 mm in carapace length. The carapace is domed, round, or oval, lacking serrations on the posterior edge. The plastron has a single hinge, a trait that will separate it from the painted turtle. Males have a red iris; females have a yellow-brown iris.

**DISTRIBUTION.** The Ornate Box Turtle is a southern Great Plains, Prairie Peninsula species with disjunct populations in the northeastern portion of its range in Wisconsin, Illinois, and Indiana. It is present in southeastern Arizona at the westernmost edge of its range. The southwest populations, including those in Arizona, have been considered a distinct species, *T. o. luteola,* and this taxon may be resurrected.

**HABITAT.** In southeastern Arizona, the species is found primarily in flat terrain and low valleys within semidesert grassland, Chihuahuan desert scrub, lower Madrean evergreen woodland, and riparian woodland with a canopy cover composed of cottonwoods and willows.

**ACTIVITY.** Plummer (2003, 2004) studied this turtle's activity patterns in southeast Arizona. He found they are above ground during the monsoon for ~90 days from July through September. Individuals are active twice during the day: in the early morning for about three hours and in the late afternoon for about 1.5 hours. They use kangaroo rat burrows for subterranean refuge. Dates for the turtle's entrance into hibernation ranged from 18 October to 6 December. Mean daily body temperatures at the beginning of hibernation were about 14–18.8°C. From December through February, body temperatures averaged about 9.8°C, slightly less than the temperature at 30 cm below the soil's surface. By April, body temperatures of turtles approached levels exhibited by surface-active turtles—a minimum of 18.8°C for moving turtles. However, turtles did not emerge from burrows until the summer monsoons began about three months later. Thus, the five-month hibernation period was contiguous with a three-month aestivation period for a total annual period of subterranean refuge of about eight months.

Figure 28. The Ornate Box Turtle. JCM

**DEMOGRAPHY.** Germano (2014) studied this turtle for 23 years at the Sevilleta National Wildlife Refuge in central New Mexico at the northern periphery of its distribution in the Chihuahuan Desert. By 2008, he found some adults 30–40 years old, and several may have been older than 40.

**REPRODUCTION.** The mean number of eggs in a clutch varied from 2.67 to 3.55. There was no indication of multiple clutches being produced, and variation in egg number was only weakly explained by the female's carapace length. Nieuwoldt-Dacanay (1997) examined females from Socorro County, New Mexico, and found evidence that clutch size is related to female body size and that spring rains may increase the proportion of females laying eggs in the summer. Turtles may defer laying eggs completely in years with dry springs rather than reducing their annual reproduction. The clutch size is one to four eggs. There is no open season for this species.

# The Mud Turtles

The Mud Turtles of the family Kinosternidae consist of 25 species in four genera. The family is restricted to the Western Hemisphere and ranges from Canada to Argentina, with the greatest diversity in Mesoamerica. The family displays many life-history traits, ranging from north-temperate to tropical latitudes and deserts to rainforests. Some species are aquatic, while others are semi-terrestrial. Females are larger in some species, but others have larger males. Mud turtles may have an extensive plastron that can completely enclose the body in the shell or a reduced plastron that exposes much soft tissue. Some species can submerge and have fully aquatic respiration, while others can aestivate underground for up to two years. Reproduction can vary from a single clutch in the spring to nesting multiple times in the summer, while others nest nearly year-round. Some have clutches of 1–2 relatively huge eggs, while others have ten or more relatively tiny eggs. Embryonic development can be direct in some species or interrupted by diapause, with incubation times ranging from 56 days to more than a year. Sex determination in mud turtles may be temperature-dependent or determined genetically by sex chromosomes.

---

## Sonoran Mud Turtle

*Kinosternon sonoriense* LeConte, 1854. (AS) Figure 29.

**IDENTIFICATION.** Adult males reach 155 mm, adult females reach 175 mm in carapace length, and hatchlings are 19–34 mm. The carapace is elongated, slightly depressed, and has three keels, with the two lateral keels disappearing with age. The ninth and tenth marginal scutes extend upward beyond the border of the other marginal scutes. The first vertebral scute rarely contacts the second marginal. The carapace is olive-brown with dark seams. The chin barbles are relatively long. The skin on the neck has a pattern of yellow reticulations.

**TAXONOMY.** Two subspecies occur in Arizona: the Sonoyta Mud Turtle, *K. s. longifemorale*, and the Sonoran Mud Turtle, *K. s. sonoriense.*

The Sonoran Mud Turtle was extirpated from California but still occurs in Arizona, New Mexico, Sonora, and northwest Chihuahua, Mexico. Yarrow (1875) described the collection of the Sonoran Mud Turtles made by Rothrock and Henshaw at Rock Creek Canyon south of Camp Apache, Arizona. In 1874, Yarrow wrote, "...they were secured while fishing in a small stream that runs through the canyon, the animals taking the bait fiercely and freely, appearing to be numerous."

The Sonoyta Mud Turtle has a short inter-anal seam and a long inter-femoral seam, about

13% of the carapace length. It has a wide first vertebral scute (28% of the carapace length) and a narrow gular scute (about 17%). The Sonoran Mud Turtle has a long inter-anal seam (19 to 23% of the carapace length) and a short inter-femoral seam (10%). The first vertebral scute is about 25% of the carapace length, and a wide gular scute is about 20% of the carapace length.

**HABITAT.** The Sonoran Mud Turtle lives in permanent waters up to 2,000 m ASL and rarely uses temporary bodies of water. When droughts occur, it will use terrestrial aestivation to survive instead of wandering overland in search of water. It can also survive in highly carbonated water—a population inhabits a collapsed travertine spring at Montezuma Well.

**DIET.** The Sonoran Mud Turtle is an opportunistic carnivore that feeds on various aquatic and semi-aquatic species and carrion.

Figure 29. Sonoran Mud Turtle. Photography JCM

**REPRODUCTION**. It matures in four to eleven years. Females lay clutches of up to eleven eggs, and the embryos may undergo diapause, a period of arrested development, and require cooling before development can start; thus, incubation may take an entire year.

## Tortoises

Desert tortoises are iconic reptiles of the Southwest. In 2002, Berry et al. noted that many populations of tortoises within the North American tortoise clade *Gopherus agassizii—G. berlandieri* complex could be designated as a species or subspecies. However, the names of the candidate taxa were unresolved. Phylogenetic relationships and the taxonomic status of desert tortoises were confused enough to prevent legal protection.

The problems associated with the identity of desert tortoise populations were examined by Murphy et al. (2011). The date of the original description of *Gopherus agassizii* was found to be 1861 rather than 1863. The species was based upon three original cotypes, only one of which exists, and they designated it as the lectotype. One cotype was destroyed in the 1906 San Francisco earthquake and subsequent fire, and the third specimen was lost. Using molecular data, the authors confirmed the lectotype is from California, not Arizona. They describe a suite of morphological characters to diagnose tortoises west and north of the Colorado River, the Mojave population, from those east and south of the river in Arizona, Sonora, and Sinaloa, Mexico. Species recognition for the southeastern population was warranted, but no names were available for the Sonoran species. Thus, the authors described and named *Gopherus morafkai* for the Sonoran Desert populations. Agassiz's Desert Tortoise, *Gopherus agassizii,* was immediately reduced to a third of its former distribution. However, the distributions are a bit more complicated than it first appears.

Captive hybrids between the Mojave Desert Tortoise and the Sonoran Desert Tortoise are impossible to distinguish from either parent species. However, non-hybrid specimens of the Mojave Desert Tortoise can be separated from the Sonoran Desert Tortoise in having a relatively narrower shell, shorter gular scutes, shorter projections of the anal scutes, and in having a flatter, pear-shaped carapace. Ecologically, the Sonoran Desert Tortoise usually inhabits valleys and alluvial fan topography, while the Mojave Desert Tortoise prefers slopes and rocky hillsides. See Appendix III for shell nomenclature.

---

## Sonoran Desert Tortoise

*Gopherus morafkai* Murphy et al, 2011 (SE) Figure 30.

**IDENTIFICATION.** Adults reach 381 mm. The carapace is high-domed and usually brown, with prominent growth lines. The plastron is yellow and has no hinge. The hind limbs are stocky and cylindrical; the forelimbs are flattened and covered with large conical scales. Males have elongated gular shields, chin glands on each side of the lower jaw, and a concave depression on the plastron. This tortoise has a narrower shell and shorter gular projections than the Mojave Desert Tortoise, but the two species are best distinguished by geography.

**DISTRIBUTION.** The species ranges east and south of the Colorado River in Arizona. In Mexico, it is present in Sonora, including Tiburon Island, and in Sinaloa, on the west side of the Sierra Madre Occidental. Introduced individuals or populations are present in North America and possibly elsewhere, but these are likely hybrids between *Gopherus morafkai* and *G. agassizii.*

**HABITAT.** The habitat is primarily the rocky slopes and bajadas of Mojave and Sonoran Desert scrub, semi-desert grassland, as well as ecotones of Sonoran Desert scrub with elements of Mojave Desert scrub and juniper woodland, interior chaparral, and desert grassland between 150 and 1,615 m ASL. Tortoises escape the temperature extremes in burrows, which are cooler in the summer and warmer in winter than ambient air temperature. Burrows are usually associated with loose soil that can be excavated. They are shallow and often located below rocks and boulders, but the tortoises may also use rock crevices, which they can't modify. Tortoises occasionally burrow under vegetation and, less often, dig burrows in soil on open slopes. They may also use caliche caves in banks of washes. Occasionally, they will rest in areas shaded by live or dead vegetation without constructing a burrow.

**ACTIVITY.** Spring activity begins as temperatures warm and decreases with increasing temperatures and drought in May and June. The time spent in burrows allows the tortoise to conserve water and energy. With the onset of the monsoon, their surface activity increases, peaking in August and September. Activity declines with the cooling temperatures of October when tortoises enter over-wintering sites. On warm winter days, tortoises may bask and forage. Females leave hibernation in late February, while some males may remain inactive throughout the spring.

**REPRODUCTION.** Eggs are laid just before or during the onset of the NAM; the eggs are fertilized by sperm stored from the previous summer's mating, and mating begins again during the monsoon. Hatchlings emerge in late summer, but some may overwinter in the nest and emerge in the spring. The mortality of eggs and juvenile tortoises is relatively high.

For many years, this tortoise was thought to obtain all the water it needed from its food,

Figure 30. A Sonoran Desert Tortoise. Photography JCM

but it has been discovered that this is not the case. Observations document tortoises drinking water after rainfall. They have been observed drinking through their nares (or nostrils) (Sullivan and Sullivan, 2016).

**DEMOGRAPHY**. Tortoises grow rapidly early in life. By five to ten years of age, they are half-grown. The growth rate slows as individuals approach sexual maturity. Tortoises reach sexual maturity when their carapace is about 220 mm (10-20 years old). Individuals may live to be 100 years of age. Males attain larger sizes than females in some populations but not in others. Adult tortoises are protected from most predators by their size and carapace. Mountain Lions, Black Bears, and humans are potential predators of adult tortoises. Population density varies from 15 to 150 adult tortoises per square mile (258 ha) (Averill-Murray and Klug 2000). There is no open season for this species. It is classified as threatened under the Endangered Species Act.

**FIELD NOTES.** The species is rare in the focus area.

# Lizards and Snakes

Squamata is the largest order of reptiles, comprising lizards and snakes, with over 11,931 species (as of May 1, 2025). They are distinguished by their scale-covered skins and movable quadrate bones, and males have hemipenes (the penis is divided into two halves). They are the sisters to the tuatara, the last surviving member of the Rhynchocephalia, with both groups belonging to the clade Lepidosauria. Most Squamata are oviparous (lay eggs), but at least 20% have evolved viviparity (live birth). Within the Squamata is a subclade that contains some lizards and all snakes – the Toxicofera includes the iguanians, anguimorphs, and snakes. Many Toxicofera can produce venom. Many squamates also have the ability to reproduce asexually but can also reproduce sexually using facultative parthenogenesis.

## Gekkota

The number of geckos currently recognized is about 1,700 species. Geckos have adapted to many habitats and lifestyles, although none are aquatic or semi-aquatic. Geckos are placed in seven distinct families that shared a common ancestor about 111 MYA, and they last shared an ancestor with all other lizards, including the Night Lizards 240 MYA. Geckos are most diverse in the tropics and the temperate regions of Eurasia, Africa, South America, and Australia. They are poorly represented in North America. Arizona has four species: three are introduced; the other is a native, nocturnal ground-dwelling species of the desert. The native Banded Gecko has movable eyelids, while the introduced Mediterranean Gecko has a single, clear scale covering the eye. The other two introduced species are not found in the focus area.

Residents in the study area are likely to call any small lizard a "gecko. " In fact there are only two species of geckos in the focus area. One is native to Arizona and can be easily recognized because it has eyelids - the Western Banded Gecko. The second species has been introduced and is easily recognized because it has no eyelids,

# The Eyelid Geckos

The Eyelid Geckos, family Eublepharidae, consists of 30 species in six genera distributed in Asia, Africa, and North America. They lack adhesive toepads (instead, they have claws) and have movable eyelids, traits absent in other families of geckos. The lineages of eublepharid and gekkonids correlate with the breakup of Pangea into Laurasia and Gondwanaland, with a recent dispersal of two African eublepharid genera from Eurasia to Africa after these two landmasses connected in the Early Miocene.

---

## Western Banded Gecko

*Coleonyx variegatus variegatus* (Baird, 1859) (SE) Figure 32.

**IDENTIFICATION.** Adults grow to 76 mm SVL, and the tail is about the same length as the body. It is a small, slender lizard with movable eyelids, vertical pupils, a thick tail constricted at the base, translucent skin with fine granular scales (tubercles absent), and long slender toes. The dorsum color and pattern are variable. It can be pink, yellow, or light grey with tan to brown transverse bands on the body and tail. Bands may be blotches in older individuals. Spurs are present at the base of the male's tail. This gecko's translucent, fragile skin will tear when the lizard is handled. In the focus area, the subspecies present is the Western Banded Gecko, *C. v. variegatus.*

**DISTRIBUTION.** The range extends from extreme southwestern New Mexico westward through southwestern Utah, southern Nevada, western and southern Arizona, and southern California and southward into Baja and Sonora, Mexico.

**HABITAT.** It inhabits arid environments, staying underground and being active at night. The Banded Gecko occurs in deserts from below sea level to 1,500 m. Banded Geckos will use the burrow of a gopher tortoise, and aggregations of geckos may use a burrow or crevice. The burrows are often located at the base of a bush.

**ACTIVITY.** From April to October this gecko is active. Body temperatures of this gecko

Figure 31. The Western Banded Gecko, *Coleonyx variegata variegata.* Photography JCM.

are relatively low; temperatures between 24 and 34°C are tolerated, but those above 40°C are lethal.

**DIET.** It feeds on various arthropods, including spiders and insects, and there is a report of one attempting to feed on a threadsnake.

**REPRODUCTION.** Females are gravid from April through September, with most carrying eggs in May and June. Hatchlings are present from July through November. Most individuals reach sexual maturity in their first year, but some do not until their second year.

**DEFENSE.** When confronted with a predator, the Banded Gecko responds quickly. The legs extend, and the body takes on a rigid posture so the body is not touching the substrate. The tail undulates slowly over the body, drawing the predator's attention to the appendage. This may allow the gecko to escape by shedding its tail and running away. Fragile skin that tears easily is another predator-defense mechanism. Should a predator grab the gecko's body, the skin will tear and allow the gecko the opportunity to escape. Predator encounters are probably frequent since 74.1% of the largest specimens had broken tails.

**MOVEMENT.** For a small lizard, Banded Geckos are mobile. Juvenile females moved an average of 10.9 m between capture points, while juvenile males moved 43.7 m. The population density at one study site was estimated to be 12 to 25 geckos per hectare. Banded geckos have specific sites they use to defecate. This may reduce predation and signal other geckos as to which individuals are present in each area.

**FIELD NOTES.** On warm nights, this lizard may be seen on blacktop roads. It carries its tail up and may be mistaken for a scorpion. They do fight with each other and I have oberved them walking away from a fight with the tail of his opponent.

## House Geckos

The family Gekkonidae has 950 described species in 51 genera. The Gekkonidae includes the house geckos (*Hemidactylus*), the Tokay Gecko (*Gekko*), and day geckos (*Phelsuma*). Species of Gekkonidae occur globally and are particularly species-rich in tropical areas. One species has been introduced into Arizona: the Mediterranean House Gecko, *Hemidactylus turcicus*. These lizards lack eyelids.

Geckos are nocturnal and are usually seen on vertical surfaces near lights. They feed on insects attracted to the lights and sometimes chase prey some distance before catching it. Females lay clutches of 1-2 eggs one to three times per year.

Invasive geckos, such as this species are considered by some relatively harmless to their new ecosystems. Yet others report significant damage to the new ecosystems. Invasive geckos can displace native gecko species, outcompeting them for food sources and altering the balance of invertebrate populations in an area, particularly when they become established in large numbers; this is considered a major ecological threat, especially on islands where native gecko species are often more vulnerable to invasion.

By consuming large quantities of insects, invasive geckos can significantly alter the insect community. Island ecosystems are particularly vulnerable: because they often have unique and endemic gecko species that are especially susceptible to the impacts of invasive geckos due to a lack of natural predators.

Invasive geckos often spread to new areas through human transportation, making them a significant concern for biodiversity conservation. Meshaka et al. (2022) suggest this Eurasian species entered North America from Key West (Florida) about 1910 and spread via cargothrough. Two populations were present in Phoneix by 1965 and it was collected in Tucson by 1970-1972.

## Mediterranean Gecko

*Hemidactylus turcicus* (Linnaeus, 1758) (i) (SE) Figure 32.

**IDENTIFICATION.** Adults grow to 60 mm SVL, with undamaged tails slightly longer than the body. The body is somewhat flattened. Skin is studded with numerous tubercles. Color and pattern are highly variable and can change rapidly. Usually, the skin is translucent pink, but it may turn white or grey to brown. The tail has transverse rings that are darker than the skin between them. Eyes lack lids and have vertical pupils. The toes expand with lamellae, which helps them cling to smooth surfaces; the toes end in claws.

**DISTRIBUTION.** Mediterranean geckos are native to southern Eurasia and North Africa and have been accidentally carried around the world as cargo.

**COMMUNICATION.** Males communicate with a squeak, telling other males to stay away from their territories. They also make a series of clicks to attract females during the breeding season.

**HABITAT.** In Arizona, they are urbanized, often living in or near buildings. However, they will go to the ground and move some distance from the building. Twice I (JCM) have found them crossing streets in an urbanized part of the focus area.

Figure 32. The Mediterranean Gecko is a non-native lizard now widespread in North America. Photography JCM

# The Beaded Lizards

In their 1956 monograph, Bogert and Del Campo diagnosed the family known as the beaded lizards, or Helodermatidae, starting with a description of their unique grooved teeth and venom glands and speculating that they are remotely related to the anguid lizards (like the Alligator Lizard discussed in this volume). The evolutionary history of the family can be traced to the Cretaceous period (145 to 166 million years ago), when the fossil species *Gobiderma pulchrum* and *Estesia mongolensis* were present. The genus *Heloderma* has existed since at least the Miocene. Beaded lizards are closely related to the monitor lizards (varanids) of Africa, Asia, and Australia. They, too, are venomous, with venom glands in the lower jaw.

The Gila Monster has four close living relatives, the beaded lizards. There are three species in Mexico and one in Guatemala.

Beaded lizards have a low metabolism and one of the lowest lizard sprint speeds. However, they have one of the highest aerobic scope values (the increase in oxygen consumption from rest to maximum metabolic exertion) among lizards, allowing it to engage in intense aerobic activity (male combat bouts) for a sustained period of time. In male-to-male combat, the dominant male lies on top of the subordinate and pins it with its front and hind limbs. While fighting, both lizards arch their bodies, pushing against each other and twisting around to gain the dominant position. The combat ends when the pinned male cannot overcome the pressure exerted by the male on top. These bouts are typically observed in the mating season. Males with greater strength and endurance are thought to gain greater reproductive success.

---

## Gila Monster

*Heloderma suspectum* Cope 1860 (AS). Figure 33-34.

**IDENTIFICATION**. The Gila Monster is the largest lizard native to North America north of the Mexican border and the most distinctive saurian in the area. Hatchling snout-to-vent lengths range from 121 to 141 mm and weigh 21-37 grams. The tail is about 20% of the body length. The largest specimens may reach 510 to 560 mm in total length. Body mass is typically in the range of 550 to 800 grams. Their heads are broad and flat, The eyes seem small for the size of the head. The head, back, and tail scales contain small, rounded osteoderms that provide protection from predators. The scales on the belly lack osteoderms.

Hatchlings have a pattern with five darkly pigmented areas on the dorsum that become obscured in their early years. Neonates have a yellow or orange-yellow and black pattern that shifts with age. Given the high contrast, the pattern is likely aposematic.

The heads of males are often larger and more triangular-shaped than in females. The length of the tail of the two sexes is statistically very similar, so it does not help differentiate the sexes. Individuals with stout tail ends occur in both nature and under human breeding. Female Gila Monsters undergo a total shed lasting about two weeks before depositing their eggs. The dorsal part is often shed in one large piece. Adult males normally shed in smaller segments in August. The young seem to be in constant shed. Adults have yellow to pink colors on a black surface. Hatchlings have a colorful banded pattern. This drastically changes

**DIET**. The diet comprises reptile eggs, bird eggs, and juvenile mammals (rodents and rabbits). Recently caught individuals have been reported to regurgitate freshly eaten, unbroken Gambel's Quail eggs. Gila monsters also appear to be feeding in part on small mammals, including young Desert Cottontails. Gila Monsters are nest predators of mammals, birds, and reptiles. This raises the question - why do heloderms need venom? Given that the venom causes significant pain when a person is bitten, it suggests Gila Monster venom is for defense instead of capturing prey (Beck, 2005).

**DISTRIBUTION and HABITAT**. It occurs in the southwestern United States and Mexico, including Sonora, Arizona, and parts of California, Nevada, Utah, and New Mexico.

Figure 33. A newly hatched Gila monster. The hatchlings have a distinct banded pattern that becomes more chaotic with age. Photography JCM

within the first six months of their lives. Hatchlings from the northern area of the species' distribution tend to retain most of their juvenile pattern. It has also been reported that Gila Monsters in areas with darker rocks and substrate will have darker colorations. Males' heads are often larger and more triangular-shaped than females. The length of the tails of the two sexes is statistically very similar, so it does not help differentiate the sexes.

Figure 34. A young Gila Monster crossing the road at night. Photography JCM

Beck (2005) noted that Gila Monsters are often associated with deserts. Yet, they avoided the most extreme deserts and used tropical thornscrub, desert grasslands, desert scrub (Sonoran and Chihuahuan), and occasionally woodlands.

**ACTIVITY.** *Heloderma suspectum* has a very sluggish metabolism, and while it can be active at any time of day (I have found them at noon and midnight) and they can be active any time of the year. Beck (2005) found it is only on the surface for about 65-70 hours of the year. During a foraging episode they may move about 215 m. Most of this activity takes place from mid-April to mid-July. During their time above ground they move about 25 km per year. One of the radio-tracked Gila monsters was observed drinking from a leaking hose in the housing area. This observation is important because it shows the lizards are using human-modified habitats to meet basic ecological needs (Nowak 2005).

# The Alligator Lizard

The Anguidae have osteoderms (a small bone) under each scale. Anguids number about 77 species in 10 genera distributed in the Western Hemisphere, Eurasia, and North Africa. They are carnivorous, terrestrial, and tend to be arboreal. Some have well-developed limbs, while others have lost all traces of external limbs and may be mistaken for snakes. The Madrean Alligator Lizard belongs to the family Anguidae and is the only representative of the family in Arizona.

There are seven species in the genus *Elgaria*. They are terrestrial, using leaf litter, burrows and crevices. They use ambush and active foraging to capture insect prey. Their activity patterns are best described as cathemeral or crepuscular.

---

## Arizona Alligator Lizard

***Elgaria kingii nobilis*** (Baird and Girard, 1852) (AS) Figures 35-36.

**IDENTIFICATION.** Adults grow to 140 mm SVL, with a tail over twice the body length. The dorsum is olive-tan, grey-olive, or beige, with eight to fifteen transverse cross bars on the body and another 20 or so on the tail. The body is covered with small smooth scales, except the scales along the vertebral line, which are keeled. The ventral scales are enlarged quadrangular plates. A trait that immediately distinguishes this lizard from all other Arizona lizards is a lateral fold with small granular scales between the dorsal scales and the ventral plates. The ventral surfaces are creamy with grey to black spots. Juveniles and adult males have broader, more vivid, black cross-bands that widen ventrally and a black mask that extends onto the face.

**DISTRIBUTION.** The species is likely present in all the Madrean Archipelago Sky Islands, ranging into the San Rafael Valley, Sonoita grasslands, and canyons of other ranges along the Mogollon Rim. It ranges southward into Mexico along the Sierra Madre Occidental as far south as Jalisco.

**HABITAT.** It has been reported in Arizona from various habitats, ranging from high-elevation coniferous forests to low-elevation desert scrub along riparian corridors.

**ACTIVITY**. These lizards are cryptozoic, often active at dusk and during the day. Like many other anguid lizards, their elongated body and reduced limbs may allow them to be momentarily confused with a snake when first observed.

**DIET.** They will forage in the leaf litter and sliding into crevices to escape capture. They feed primarily on arthropods, but large individuals may prey on small vertebrates.

**REPRODUCTION.** Mating occurs in the fall, and eggs are laid in the spring. Females lay clutches of 9 to 15 eggs and may attend the eggs until they hatch. The Madrean Alligator Lizard can be distinguished from the skinks by its greatly elongated body and the fold of granular scales separating the dorsal and ventral scales.

Figure 35 Madrean Alligator Lizards (top) Photography JCM Figure 36 (bottom) Devon Medler and Ben Saldiva

The bold pattern of juveniles and adult males is likely to be aposematic, warning predators not to eat them -its unclear if the warning is based upon some unknown chemical defense that the lizard has, or if the lizards are mimicing another species.

**TAXONOMY.** The subspecies present in Arizona is the Arizona Alligator Lizard, *E. k. nobilis* Baird and Girard.

# Collard Lizards & Leopard Lizards

The dozen lizard species in the family Crotaphytidae are in two genera *Crotaphytus* and *Gambelia.* They are species that stalk and run-down small lizard species. Their distribution is restricted to western North America.

Collared and Leopard Lizards were once placed in the Iguanidae but are now in the iguanids sister family, Crotaphytidae. The oldest fossils of Collared Lizards and Leopard Lizards are from two localities in Lincoln County, Nevada: a fossil maxilla of a collared lizard and a dentary bone from a Leopard Lizard. The two lineages are easily separated by their teeth. Collared Lizards (*Crotaphytus*) have broad, robust teeth, most of which have three cusps and show few or no recurved teeth. Leopard Lizards (*Gambelia*) have recurved teeth throughout the tooth row, which contains more single cusp than tricuspid teeth. The two genera can also be distinguished by the rows of scales that separate supraorbital semicircle scales. The Collared Lizards have one or two rows of large scales separating them, while the Leopard Lizards have several rows of small scales separating them. The published records of valid Collared- and Leopard Lizard fossils date to the middle Pliocene (4.18 to 3.58 MYA). However, at least one molecular study suggests that the two genera diverged about 25 MYA (Townsend et al., 2011).

The nine species of collard lizards have dramically smaller home ranges than there leopard lizard relatives. Rodda (2020) found collard lizards have a home range that is a mere 0.002 % of the leopard lizards home ranges. Collard Lizards readily break their tails to esxcape predators, and hide from predators in the vegetation. In the study area the Eastern Collard Lizards is uncommon to rare. There are litteraly a few musuem specimens and one record of the species on iNaturalist.

There are two recognized species in the leopard lizard genus *Gambelia*. A report by Claunch and Taylor (2017) documents this species killing a Desert Iguana and only eating the tail, which is unusual because they often swallow the entire animal.

---

## Eastern Collared Lizard

*Crotaphytus collaris collaris* (Say, 1822) (IM) Figures 37-38.

Figure 37. Eastern Collard Lizard.
Photography JCM

**IDENTIFICATION.** Adults reach 116 mm SVL; the total length can be 356 mm. Hatchlings are about 90 mm in total length. The dorsum may be green, brown, blue, or olive, depending on the sex, age, and substrate background color. There are two dark collars; the first does not extend around the ventral side of the throat; the second may reach the front legs, particularly in males. The Sonoran Collared Lizard has a complete anterior collar on the ventral side of the throat. The throat is spotted, and the tail is round, with spots and transverse bands. Females often have a uniform or slightly spotted gular region. The feet and head may also sport a striking yellow-to-orange color. These colors make them among the more attractive species for lizard-watching tours.

**DISTRIBUTION**. The species ranges from western Illinois and Missouri westward to Utah, Nevada, and Arizona and southward to Baja California, Sonora, Chihuahua, Durango, Coahuila, Zacatecas, and Nuevo Leon.

Figure 38. Eastern Collard Lizard. Photography JCM

**HABITAT.** The Eastern Collard Lizard uses a variety of habitats, ranging from grasslands to deserts and riparian woodlands from sea level to 2,400 m. They are always found near rocks, especially boulders, where they can bask in the sun to warm up, watch for predators and prey, and watch for other collared lizards.

**DIET.** Their diet is omnivorous but heavily dependent on insects and smaller lizards. Best and Pfaffenberger (1987) found 63 categories of arthropods, as well as eight lizards and one snake in the diet of a New Mexico population of this lizard.

**DISPLAYS.** Male Eastern Collared lizards pay a price for their bright color and the displays they make to attract females and defend territory. Males are the subject of more attacks by predators than are the less brightly-colored females. Using plasticine models of Eastern Collared Lizards, Husak et al. (2006) found models exhibiting a strong color contrast with rocks detected and attacked most often. By comparison, inconspicuous models of females were never attacked at any of the study sites.

Female Eastern Collared Lizards display an orange coloration on the sides of the body with the production of the first clutch of eggs. The orange marks are maintained throughout the season. Females use the orange patches as a signal to males. Experimental observations of lizards revealed that males court females with orange markings more intensely than females who have not yet developed pigmentation.

**REPRODUCTION.** Courtship becomes more frequent within 24 hours after females have had orange paint applied to their bodies. Females living within a territory controlled by a male are most likely to mate with that male first. Juvenile males also have an orange dorsolateral coloration that resembles gravid females. The idea that juvenile males mimic females to reduce aggression from adult males has been examined, and the results have been inconclusive.

Females may lay one or two clutches of eggs per season, with clutch sizes ranging from 1 to 14. Santoyo-Brito et al. (2017) incubated eggs of this species at one of seven constant temperatures. More males were produced at lower and higher temperatures and more females at intermediate temperatures. Although none of the seven treatments produced only females or only males, incubation at different constant temperatures influenced hatchling sex ratios.

**FIELD NOTES.** This lizard is uncommon in the study area. I have included it here based on museum specimens and iNaturalist data. The Desert Spiny Lizard and Clark's Spiny Lizard both have collar-like markings and may be mistaken for this species. Collared lizards have relatively small, smooth scales compared to the spiny lizards, with large overlapping, keeled scales.

---

## Long-Nosed Leopard Lizard

*Gambelia wislizenii* (Baird & Girard, 1852) (SE) Figures 39-40.

**IDENTIFICATION.** Adult females reach 144 mm SVL, and adult males reach 119 mm SVL. Hatchlings are 100 to 130 mm in total length. This is a large, tan lizard with a series of dark brown spots dorsally and white transverse bands that connect on the sides to form cross bands. Like the collared lizards, females develop bright orange bars or spots on their sides. Juveniles have red spots that later become dark brown spots. Their body form is like that of collared lizards, with a muscular head (although the head is not as robust as that of Crotaphytus) and a long tail. The species has several rows of small scales between the supraorbital semicircle scales.A pattern with white spots is present in lizards using sandy habitats.

**DISTRIBUTION.** The species ranges from southeast Oregon and southern Idaho southward through Nevada, Utah, eastern California, Arizona, New Mexico, and Texas; southward into Mexico, including northeastern Baja California, western Sonora, Mexico, and north-central Mexico.

**HABITAT.** They occur primarily in desert habitats but also use adjacent vegetation types, including semi-desert grasslands and pinyon-juniper woodlands. They are one of the few lizards inhabiting all four deserts in the United States.

They prefer open areas where they may run. Like collared lizards, they are visually oriented, sit-and-wait predators that are very tolerant of high temperatures. However, unlike Crotaphytus, they tend to bask on cobble-sized rocks, berms, or open terrain where they can chase prey. Leopard lizards are common in areas of creosote bush, sagebrush, and even dunes stabilized with vegetation.

**DIET.** Prey are mostly other lizards, supplemented with arthropods. Leopard Lizards seek shelter in burrows of small rodents or the Desert Tortoise to regulate their body heat, and gravid females may use the burrow to deposit their eggs (Walde et al., 2015).

**ACTIVITY.** These are highly mobile lizards that use a different burrow each night. The

Figure 39. Above. Long-Nosed Leopard Lizard. Photography JCM

Figure 40. Below. A juvenile Leopard Lizard. Photography by Lee Doolan

males are often tolerant of one another, with only a third defending their space. They are associated with dunes and are present in riparian habitats. Because they live indrier habitats with less photosynthesis they require larger home ranges than collard lizards (Rodda, 2020).

**REPRODUCTION**. Females may lay up to two clutches annually at southern latitudes, although often only one clutch is laid annually. Eggs are laid in a burrow and incubated for five to seven weeks. In August, when adult activity is winding down, the young appear. Leopard lizards have relatively few, hatchlings. The appearance of juveniles coincides with the end of seasonal adult activity. Juvenile coloration mimics gravid females and it may reduce cannibalism.

## North American Spiny Lizards

The North American Spiny Lizards, family Phrynosomatidae comprises ten genera and 150 recognized species that range from southern Canada to Panama. It reaches its most remarkable diversity in the southwestern United States and Mexico, where they are the most conspicuous part of the herpetofauna. Most are small to medium in size, usually less than 100 mm, but the largest species reach about 150 mm. Phrynosomatids are insectivorous, diurnal lizards that exploit a variety of terrestrial environments. Some specialize in rocky habitats; others inhabit dunes and sand flats, yet others are tree-dwelling or use grassland habitats.

They occur below sea level in California (i.e., Death Valley) to the highest peaks in southern Arizona. There are numerous species, and I discuss them by clades. Arizona has eight of the ten genera and about 37 species and subspecies.

## Sand Lizard Clade

The sand lizard clade includes the genera *Callisaurus, Cophosaurus, Holbrookia,* and *Uma.* These lizards are adapted for living in loose soils, from loamy and gravel substrates to dunes. Some bury themselves in sand to escape predators or high surface temperatures. Callisaurus and Cophosaurus are sister species and ecological counterparts, with the former found in all western deserts and the latter primarily in the Chihuahuan Desert and the adjacent Arizona upland Sonoran Desert. In some areas of Arizona, the focus area and the ranges overlap.

*Holbrookia* are small, earless sand lizards with oval bodies, found in the Chihuahuan and adjacent Sonoran Deserts, as well as the Colorado Plateau. They are poorly understood taxonomically, with names fluctuating from one paper to another. The taxonomy of Arizona's *Holbrookia* has been controversial because the lizards display a confusing array of color and morphological variations. Two species of *Holbrookia* are recognized in Arizona by Holycross et al. (2024). That arrangement is followed here. *Uma* are the most specialized sand lizards, and do not occur in the area covered here.

---

### Zebra-tailed Lizard

*Callisaurus draconoides ventralis* (Hallowell, 1852) (SE) Figure 41-42.

**IDENTIFICATION.** Adults reach a maximum SVL of 101 mm. Hatchlings have a SVL of 26 to 38 mm. Males mature at about 70 mm SVL; females mature at about 63 mm SVL. Males are larger than females. The dorsum is generally tan in color and matches the desert substrates. There are paired spots along the back. Males have a pronounced set of black bars on each side of the axilla, set against a blue ground color. This is visible on the underside or from the side, especially if you are at the lizard's level. Females and juveniles can have bars on the sides and the ventral, but it is much less pronounced. This species has external ear openings, unlike the superficially similar Greater Earless Lizard. Both species exhibit zebra-like black and white markings on the underside of their tails. The Greater Earless Lizard also has ventral black bars set more towards the rear legs than the forelegs.

**DISTRIBUTION.** The species occurs in the Great Basin, Mojave, and Sonoran Deserts and occupies open habitats. Creosote bush flats are an excellent place to find them, an environment where they have plenty of room to stretch their long legs. They are ubiquitous in washes, which often have loose soil.

**ESCAPE BEHAVIOR.** When first seen, they appear as a fast-moving shadow that crosses your path, stops suddenly, and waves its tail at you. (The Greater Earless also does this.) Blink, and you may lose the location of the lizard as it runs off a dozen meters before another sudden stop. Zebra-tailed Lizards are the speedsters of the deserts. They can bury themselves below the surface in loose sand by rapidly vibrating their body from side to side. This allows them to wait for prey and avoid predators. Both sexes seem territorial and display head bobbing, expanding dewlaps, and push-ups.The waving tail behavior may signal the predator that it has been spotted and that a continued chase is pointless.

**SPEED.** Some of the longest limbs seen on any North American lizard are found on Zebra-Tailed Lizards, a trait correlated with open habitats. When sprinting, they can attain speeds of 28 km/hour.

Figure 41-42. Zebra-Tailed Lizard . Top Adult Bottom subadult. Photography JCM

**DIET** Several studies suggest that caterpillars, beetles, and wasps are important in the diet of this species. But it seems likely that they eat a greater variety of insects than those listed in the literature.

**REPRODUCTION**. Hatchlings are present from mid-July through November and grow (6.5 mm/month), with most individuals attaining minimum adult size at less than a year old. They emerge from the egg, ready to perform their tail-waving display. Home ranges or territories are relatively small, with the mean distance between captures for juveniles 20.3 m and 31.9 m for adults (Pianka and Parker, 1972).

**PREDATORS.** include a variety of other lizards, snakes, birds, and mammals.

**TAXONOMY.** Two subspecies occur in Arizona: the Western Zebra-Tailed Lizard, *C. d. rhodostictus* and the Eastern Zebra-Tailed Lizard, *C. d. ventralis*. These are very similar in appearance. The former is found along the western edge of Arizona and extends into California. It occupies the Sonoran, Mojave, and Great Basin deserts. The Eastern Zebra-Tailed Lizard is found primarily in the Sonoran Desert, east of the influence of the Colorado River.

---

## Chihuahuan Greater Earless Lizard

*Cophosaurus texanus* Troschel, 1852 (SE) Figure 43.

**IDENTIFICATION.** Adult males grow to 89 mm SVL, adult females to 78 mm; hatchlings are about 38 mm SVL. This lizard has no external ear openings, its legs are relatively long, and vertebral scales are slightly larger than lateral scales. It has seven-keeled upper labials, and a gular fold is present. The dorsum is grey, with two rows of longitudinal spots that extend onto the tail.

This species is remarkably similar to the Zebra-Tailed Lizard, so much so that they could be considered mimics, except that their ranges overlap very little. The absence of an ear opening distinguishes this species from the Zebra-Tailed Lizard. The position of the black bars on the sides is even easier to see with the naked eye. In this species, the bars are closer to the hind legs, while in the Zebra-Tailed Lizard, they are closer to the front limbs. Sexual dimorphism is apparent, especially during the breeding season when the males can take on vivid coloration. The bars on the side are more vivid, as in Zebra-Tailed Lizards, but they also get bright green flanks and orange spots dorsally. This is one of the few species of lizards in which males produce an orange coloration, generally reserved for gravid females. Females turn orange when gravid, but the color pattern is more of a wash than spots.

**DISTRIBUTION.** In Arizona, the subspecies are the Chihuahuan Greater Earless Lizard, *C. t. scitulus* (Peters, 1951). This is primarily a Chihuahuan Desert species, but its range goes well west into Arizona, almost reaching the Colorado River. When populations occur in the vicinity of Zebra-Tailed Lizards, the two species partition their habitat, with Zebra-tails being the low-desert adapted form.

**ACTIVITY.** The Chihuahuan Greater Earless Lizards are active in the spring through fall, when temperatures are conducive to surface activity. They use elevated terrain to attract attention and broadcast their presence to members of their species. They can often be seen doing push-ups, especially during the breeding season. Habitats used are sometimes open, flat, sandy, or gravelly areas of desert and bunchgrass habitats, but they also use adjacent, rocky slopes. In Arizona, they can be very abundant in arroyos. Like *Callisaurus,* they wave their tail displaying the black bars on the underside,

---

## Elegant Earless Lizard

*Holbrookia elegans* Bocourt 1874 (VU) Figures 44-47.

**IDENTIFICATION.** Adults are 54 to 75 mm SVL, with a tail equal to the body length of females and longer than the body length of males. Typically Elegant Earless Lizards have a pattern of chevrons on the back, which is more pronounced in females; males tend to be more spotted and have a more distinct set of black bars behind the front legs, like other *Holbrookia* and its larger cousin, the Greater Earless Lizard. All *Holbrookia* lack the black bars on the underside of the tail. This is a long tailed species, a characteristic that can be

Figure 43. A male Chihuahuan Greater Earless Lizard. Photography JCM

used to differentiate it from *H. maculata* in southeast Arizona.Tail length in males is longer than the body length, while the tail length in females is equal to or slightly shorted than the body length.

Gravid females are distinctive, with an orange spot on the throat and an orange wash over the forebody. Males may have a faint orange spot and blue under the black lateral bars next to the upper abdomen, which is most pronounced during the breeding season.

Some populations of the Elegant Earless Lizard have a plain morph (most common in the Huachuca and Patagonia mountains) that lacks dark dorsal markings. However, the black bars are present, especially in males. This species also has circumorbital scales similar in size to the supraoculars, with eight to ten chevrons on the tail. Gravid females have a red-orange spot on the throat. Figure 44-47 shows some of the variation in the populations found in the focus area.

Figure 44-47 shows some of the color and pattern variation in Elegant Earless Lizard. The plate is organized from upper left and moves from left to right

**HABITAT.** Inhabits semidesert grassland and Madrean woodlands and perfers areas with sparse vegetation.

**REPRODUCTION.** During the active season, March through October, sexual pairs form. After the monsoon rains, females lay 6–12 eggs. This species is most active on sunny, warm days. (See the taxonomic comment in the introduction to Sand Lizards.)

**TAXONOMY.** The genus *Holbrookia* is a taxonomic mess, particularly *H. elegans* and *H. maculata*. They have been considered two species with subspecies as well as one species, *H. maculata* with *elegans* as a subspecies.

**FIELD NOTES.** Often seen sitting on top of small rocks a few inches off the ground. Its frequently in colonies.

## Horned Lizards

Horned lizards (genus *Phrynosoma*) have evolved novel adaptations not seen in other lizards. Their flat body, covered with spikes of dermal armor, the longest on the back of the head, makes them instantly recognizable. Arizona has seven of the 22 extant species.

The molecular clock estimates horned lizards evolved about 30 MYA (Leache and Linkem, 2015). The variety of adaptations to protect themselves from environmental hazards is spectacular. Horned lizards are masters of camouflage. By remaining still and matching their coloration to local substrates, they avoid detection by many visually oriented predators. Should that fail, they can run to a burrow and inflate their body to convince a predator they are too large to be swallowed. And, their spiny armor makes difficult to swallow. They use the intricate arrangement of the scales on their back to harvest water from raindrops, directing the water on the lizard's body to the corners of their mouth. Perhaps the most novel adaptation is the ability to squirt blood out of their eyes in a defensive gesture, at least in some species. Squirting blood from the eyes was first reported in the *Proceedings of the Zoological Society of London* in 1871. Famed naturalist Alfred Russell Wallace read the following parts of a letter received from John Wallace of Stockton, California, to the Society:

"Under certain circumstances, apparently as a means of defense, this creature squirts out from one of its eyes a jet of bright red liquid very much like blood. This I have observed three times from three different individuals, although I have caught many that did not do it. They do not generally use this defense when first captured, although I caught one a few days ago which squirted the liquid a distance of 6 inches over the back of my hand, and another ejected it when I flourished a bright knife before its eyes."

Horned lizards feed heavily on ants, especially harvester ants, which include some of the most venomous insects. Several species, including the Texas Horned Lizard, *Phrynosoma cornutum*, and the Regal Horned Lizard, *Phrynosoma solare*, may accumulate formic acid or another toxic molecule from their ant prey in their bloodstream, then spray their toxic, noxious, irritating blood from their eyes into the faces of coyotes and foxes as a deterrent. Ant-eating in horned lizards appears to be associated with a shortening and sliming of the mandible, a reduction in the total number of teeth, a decrease in head height, and a reduction in the area of insertion of the jaw adductor musculature in species that have a higher percentage of ants in their diet (Meyers et al. 2006).

Horned lizards obtain water from food, conserve water by staying underground, and can use the scales on their body to collect rainwater for drinking during a storm. They also regulate their water balance by excreting excess salt through salt glands in their nostrils. Most female horned lizards lay eggs in nests they excavate. Larger females tend to produce more eggs than smaller females. However, the Greater Short-horned Lizard is an exception, as discussed below.

## Greater Short-horned Lizard

*Phrynosoma hernandesi* Girard, 1858 (nd) Figure 48-49.

**IDENTIFICATION**. Adults reach 112 mm SVL, with a tail less than half the body length. Hatchlings are 25 to 32 mm SVL. Females are larger than males. Short horns, smooth ventrals, and a single row of enlarged ventrolateral fringe scales distinguish this species from other species. The dorsum is variable and will usually match the substrate. It ranges from tan or brown to yellow or red. Usually, there is a pair of dark neck blotches, followed by a transverse dark blotch that crosses the vertebral line at mid-body; these blotches follow and extend onto the tail. A vertebral stripe is sometimes present. The belly is mottled grey and white. The very short, widely separated occipital horns will distinguish this species from all other horned lizards, which have long, sharply pointed occipital horns.

**DISTRIBUTION**. The Short-horned Lizard ranges from Alberta and Saskatchewan, Canada, southward through the United States to Durango, Mexico. It uses a variety of habitats, ranging from short grasses and shrub communities to oak-juniper woodland along the lower slopes of mountain ranges and at higher elevations in Arizona, pine forest. Greater Short-Horned Lizard populations occur in open areas with friable soils where the trees and shrubs are widely spaced. It can be found from 600 to 3,200 m ASL. Hernandez's Short-horned Lizard, *Phrynosoma hernandesi hernandesi,* is found over most of Arizona, with the Colorado Plateau in the northeast being a zone of intergradation with the New Mexico Short-Horned Lizard *P. h. ornatissimum.*

High elevation populations have a short season to find food and reproduce. Therefore, males become sexually mature after their first year of life, while females generally take two hibernation periods to mature. Mating occurs in May and June.

**REPRODUCTION.** Females are viviparous. Litters of up to 48 young are born from July to September. The neonates are small, weighing about one gram, have no horns, and cannot crawl until a day old.

Figure 48. Greater Short-horned Lizard. Photography Robin Agarval

Figure 49. Neonate Greater Short-horned Lizards Photography JCM

## Regal Horned Lizard

*Phrynosoma solare* **Grey, 1845 (SE) Figures 50-52.**

**IDENTIFICATION.** Adults reach 117 mm SVL. The dorsum is tan to grey with a sizeable light-colored area centered on the back and bordered with darker pigment. The light area has indistinct cross bands that extend onto the tail. A vertebral stripe may be present. They have four large horns with their bases in contact (a trait found in no other Arizona horned lizard) and a single row of enlarged fringed scales bordering the edge of the body.

**DISTRIBUTION and HABITAT**. The species ranges from central Arizona southward to Sinaloa and Baja, Mexico. It occurs in Arizona upland desert scrub, Lower Colorado Desert Scrub, Chihuahuan Desert Scrub, Semi-desert Grasslands, and Coastal Scrub in Baja. It uses flats with pebbles and rocks and ranges from sea level to 1,400 m. Parker (1975) found the Regal Horned Lizard in five different microhabitats at a Maricopa County study site: desert scrub dominated by creosote and bursage; a disturbed desert riparian community; on a north-facing slope with bursage, brittlebush, paloverde, and staghorn cholla; on a south-facing slope with teddy bear cholla; and the valley floor with bursage, brittle bush, paloverde, ironwood, and boulder piles. More lizards were found on the valley floor than in any other microhabitat. Activity was between 9:00 a.m. and 11:00 a.m. from October to April and between 6:00 a.m. and 9:00 a.m. for the remainder of the year. However, this lizard may also be active when temperatures are favorable at dusk or bask on asphalt roads.

**DIET.** Like other horned lizards, harvester ants compose much of its diet. These noxious, toxic ants are coated with a mucous layer before swallowing. The lizard's blood chemistry is capable of neutralizing the toxins.

**REPRODUCTION.** Courtship and mating occur in spring, and females deposit up to 33 eggs in July and August; the eggs hatch in 5 to 9 weeks. Hatchlings are 26 to 31 mm SVL. The young grow rapidly and can add 15 to 20 mm in the fall before hibernation.

**PREDATORS**. Predators include carnivorous mammals, birds, other lizards, and snakes. However, the horns and spike-like scales make this species difficult to swallow, and predators have died while trying to make a meal of it. The species has a repertoire of defensive behaviors, which include freezing in place, inflating its body, running away, and squirting blood from its eyes.

Figures 50-52. Regal Horned Lizard.Top and middle are adults, Bottom is two Regal Horned Lizards in copula. Photography JCM

# Spiny Lizards

The 114 species in the genus *Sceloporus* are some of the most frequently observed lizards in the Western Hemisphere. While they are primarily insectivorous, large specimens of *S. magister*, have been reported to eat small birds. *Sceloporus* tend to be diurnal, basking to warm their bodies, yet one *S. clarkii* that sometimes visits my backyard can be found feeding on insects attracted to lights on warm summer nights. Most *Sceloporus* are oviparous, but high-altitude species like *S. jarrovii* have evolved viviparity. Few field guides in the last 20 years have recognized *Sceloporus bimaculosus* as a valid species. Often, it seems to be considered a subspecies or pattern morph of *Sceloporus magister*. It is not common in the focus area, but it lives alongside *S. magister*.

---

## Twin-Spotted Spiny Lizard

*Sceloporus bimaculosus* Phelan & Brattstrom, 1955 Figure 53.

**IDENTIFICATION**, A large, spiny lizard that reaches a body length of 119 mm, with two rows of spots on the dorsum of the body and wedge-shaped black patches on the shoulders that may form a complete collar. The Twin-spotted Spiny Lizard can be distinguished from Clark's Spiny Lizard (*Sceloporus clarkii*) by dark cross-bars on the front limbs and three or four scales that extend backwards over the ear openings. The Desert Spiny Lizard (*S. magister*) is very similar to the Twin-spotted Spiny Lizard. The only trait that distinguishes the two is the dorsal pattern. Two rows of dark paravertebral spots, in bimaculosus verses a dark or purple mid-dorsal stripe or blotch 4-5 scales wide in magister.

**DISTRIBUTION**. This is a species of the Chihuahua Desert. The species ranges from eastern Arizona, western New Mexico, and western Texas southward into Mexico to central Chihuahua, southern Coahuila, and northeastern Durango.

**DIET.** Omnivorous, on invertebrates, occasionally small lizards, flowers, and berries.

**REPRODUCTION**. Females mature at about 80 mm and can be gravid between April and August. Clutches contain 4 to 10 eggs, and hatchlings are about 30 mm SVL.

**TAXONOMY**. *Sceloporus magister bimaculosus* was elevated to full species by Schulte et al. (2006) but returned to the synonymy of *S. magister* by Leaché and Mulcahy (2007). Recent authors have alternatively treated it as a full species or a synonym (e.g., Laczano et al. 2019, in whose list it does not occur in Coahuila). Pavón-Vázquez et al. 2024 do not recognize *bimaculosus* as a valid species or even subspecies.

Figure 53. Twin-spotted Spiny Lizard. Photography JCM

## Clark's Spiny Lizard

*Sceloporus clarkii* Baird & Girard, 1852 (SE) Figure 54.

**IDENTIFICATION**. Adult males reach 144 mm SVL; adult females to 120 mm SVL. The dorsum is covered with keeled, spine-tipped scales. The forearms have transverse bands that distinguish this species from the Desert Spiny Lizard. The coloration is quite variable and changes with temperature. The upper surfaces of the body can be grey, tan, or blue, with scattered turquoise scales mixed with tan and brown. They have a dark collar under or around the neck; males have vivid blue throats and ventral blue patches. Females develop orange to red heads during the breeding season.

**DISTRIBUTION**. This abundant species occurs in central and southeastern Arizona from the lower portion of the Mogollon Rim in north-central Arizona and the Mogollon Plateau of New Mexico southward along the Sierra Madre Occidental to Jalisco, Mexico. The elevational range extends from sea level to about 1,848 m ASL.

**HABITAT.** It uses biomes ranging from Arizona upland Sonoran Desert Scrub to Great Basin Conifer Woodland. Clark's Spiny Lizard occurs primarily in wooded areas of oak-pine woodland, tropical deciduous forest, subtropical thorn forest, gallery woodland communities that extend into more arid habitats, rocky areas at the lower edges of foothills, and oak-grassland communities as low as 1,060 m. It is primarily arboreal and spends time in the canopy; it also frequents rock outcrops near the ground. Woodrat nests in rock crevices and trees are used as refugia and probably as foraging sites.

**DIET.** This species is omnivorous, eating a variety of arthropods. Beetles, ants, wasps, grasshoppers, and spiders are important in the diet, as are leaves, buds, and flowers. This normally diurnal lizard will come to lights on buildings at night to feed on the insects they attract.

**BEHAVIOR**. Clark's Spiny Lizards tend to perch very high in trees and are only observed

Figure 54. Clark's Spiny Lizard. Photography JCM

while basking lower on tree trunks. This lizard's usual escape behavior is to run up a tree or move to the opposite side of the tree or rock it is on. It has also been seen jumping into a stream and submerging itself to escape a human. Selection for color matching may be stronger in the treetops, where they are more visible to overhead avian predators. Proctor Trail in Madera Canyon in late afternoon is an excellent place to observe this species.

**REPRODUCTION.** Females are sexually mature at 89 mm SVL. Clutch sizes are 7 to 28 eggs, with larger females producing larger clutches.

**TAXONOMY.** Two subspecies are recognized in Arizona: the Sonoran Spiny Lizard, *Sceloporus clarkii clarkii* (the Sonoran Desert population), and the Plateau Spiny Lizard, *Sce-*loporus clarkii vallaris (the Colorado Plateau population).

**FIELD NOTES.** Proctor Trail in Madera Canyon in late afternoon is an excellent place to observe this species.

---

## Southwestern Fence Lizard

*Sceloporus cowlesi* Lowe & Norris, 1956 (SE) Figure 55.

**IDENTIFICATION.** Adult males reach 75 mm SVL; females are slightly larger. The dorsum has heavily keeled, imbricate scales with a spiny tip. It is a brown, tan, or black (at low temperatures) lizard with a pale dorsolateral stripe and a pale ventrolateral stripe. Adult males have patches of intense blue on the throat and venter, but these are absent or faint in females and juveniles. This species is easily confused with other small *Sceloporus*, particularly the Plateau Lizard, Striped Plateau Lizard, and Slevin's Bunchgrass Lizard. It was long considered a subspecies of the Eastern Fence Lizard (*Sceloporus undulatus*), and some authors still refer to it as belonging to that species. Slevin's Bunchgrass Lizard has lateral scale rows parallel to the body's axis. The Striped Plateau Lizard has light dorsolateral stripes absent in this species. The Plateau Lizard and this species are morphologically cryptic and are not diagnosable based on morphology.

**DISTRIBUTION.** The Southwestern Fence Lizard is found in much of southeastern Arizona and east-central Arizona at elevations between 1,000 and 3,000 m. It inhabits various communities, from Chihuahuan desert scrub to Madrean evergreen woodland and montane coniferous forest.However, it appears absent from the Santa Ritas and Madera Canyon. It is present in the Las Cienegas Wetlands, in the vicinity of Patagonia, Rio Rico, and Tubac

**DIET.** The prey are primarily arthropods, including ants, beetles, grasshoppers, and flies.

**BEHAVIOR.** Bright blue male markings communicate information to other lizards of the same species. However, sexual confusion has been documented in Southwestern Fence Lizards. Dittmer et al. (2015) found that signal patch morphology between the sexes seems confused. They compared the morphology of signal patches between males and females in three species of related spiny lizards. The results suggest that the similar patch size between the sexes of both *S. cowlesi* and *S. tristichus* causes sexual confusion.

**REPRODUCTION.** Females lay clutches of up to 12 eggs in the spring and summer. Bateman and MacCoubrey (2012) studied this lizard over seven years. They found hatchlings in May, and their activity peaked in September. Adult activity peaked in early June. Hatchlings grow rapidly: most reached 50 mm, and a few attained 60 mm body lengths at the end of their first summer. Some individuals were captured over four or five summers, but most were not seen after their third summer.

Figure 55. Southwestern Fence Lizard. Photography Michael Andersen

## Yarrow's Spiny Lizard

*Sceloporus jarrovii* Cope, in Yarrow, 1875 (SE). Figure 56.

**IDENTIFICATION**. Adults reach 106 mm SVL, with hatchlings 23 to 32 mm SVL. The head may be flecked or almost all black, and the neck has a collar with the posterior edge lined with white. The dorsum is covered with black-edged scales, forming a mesh-like pattern that gives the lizard the bronze, green, blue, or pink cast. Males have blue throat and belly patches. In females, the blue is subdued, often grey or black. This species has no transverse bars on the forelimbs, and the supraocular scales are in a single row: combining these traits will separate it from all other Arizona *Sceloporus*.

**DISTRIBUTION and HABITAT.** This Sky Island species inhabits the Madrean Evergreen Woodland, Montane Conifer Forest, and Subalpine Conifer Forest communities from 1,100 to 3,000 m in elevation. It is usually associated with rock habitats, such as outcrops, boulders, cliffs, and talus slopes. It occurs in southeastern Arizona, southwest New Mexico, and central Mexico.

**DIET**. Yarrow's Spiny Lizard feeds on insects, spiders, and plant matter. Stomach content analyses found ants, termites, beetles, true bugs, and lepidopterans. Neonates feed predominantly on small true bugs and ants.

**REPRODUCTION**. Females are sexually mature at 50 mm SVL, and males at 46 mm SVL. Unlike most other spiny lizards, this species is viviparous. Studies of its placenta suggest the female can pass molecules to the embryos from her bloodstream. The evolution of live birth in these lizards is an adaptation to living in cool, high-altitude environments, where the female can move the embryos to suitable temperatures for development (Anderson et al., 2011).

**ACTIVITY.** Each of the Sky Island populations tends to have distinct coloration and patterns. Populations of this lizard may exceed 200 individuals per hectare. During the summer, they defend territories but abandon them to feed on swarming insects. They also have communal overwintering sites and may overwinter in groups of 60 or more. On warm winter or spring days, a single rock may be covered with dozens of these lizards. However, during the summer, these lizards are territorial, and they chase other lizards and do push-up displays.

Figure 56. Yarrow's Spiny Lizard. Photography JCM

## Desert Spiny Lizard

*Sceloporus magister* Hallowell, 1854 (SE) Figure 57.

**IDENTIFICATION.** Adult males reach 142 mm SVL, with the tail slightly longer than ward to southern California, northern Baja Mexico, Arizona, and southwest New Mexico southward to Mexico's west coast. Males have a dark blue stripe (or blotch) on their dorsum.

**HABITAT.** Arizona covers much of the state and ranges from near sea level to about 1,700 m. It inhabits desert flats, semi-arid grasslands, low mountain slopes, riparian woodl-lands, and flat terrain with creosote bush, mesquite, yucca, and grasses.

It occurs in various biomes, from Sonoran Desert scrub to Interior Chaparral. It is on the ground almost as often as it is in trees. This species can be terrestrial, arboreal, and saxicolous (living among rocks). It will not hesitate to enter a mammal burrow or run up a tree to escape. It is often associated with cactus clumps and may hide in the base of the plant or climb into its spiny stems to avoid predators. Desert Spiny Lizards adapt to urban areas with low-density housing and often inhabit gardens and yards. They are associated with packrat nests.

**DIET.** Omnivorous. They consume a variety of arthropods and eat smaller prey than many other species of similar size. Large adults may eat nestling birds, bats, fruits, flowers, and leaves.

**REPRODUCTION.** Females are mature at about 80 mm; males mature at about 95 mm in body length. The average clutch size is 8.4 eggs, but they may produce as many as 19 eggs.

Figure 57. Desert Spiny Lizard. Photography JCM

## Slevin's Bunchgrass Lizard

*Sceloporus slevini* Smith, 1937 (IM) Figure 58.

**IDENTIFICATION.** Adult males reach 59-68 mm SVL, with females slightly larger, reaching 70 mm. The hatchlings are minute, about 20 mm long, and weigh only 0.3 grams.

The dorsum is brown or tan, with crescent-shaped dark spots on either side of the vertebral line. A pale dorsolateral stripe extends from the head to the tail's base. A blue spot on the shoulder is bordered by black (in males only). The gular region is uniform. The tail has a dark brown or black midline stripe. Males have an orange ventrolateral stripe bordered medially by a blue patch. The belly is a uniform cream or white. Some populations have a patternless morph in both sexes. Femoral pores (glands) on the underside of the femur from the right and left sides are separated by one or two scales; in other species of Sceloporus, they are separated by three or more scales.

**DISTRIBUTION and HABITAT.** The species occurs in southeastern Arizona (the Chiricahua, Huachuca, and Santa Rita Mountains) and adjacent extreme southwestern New Mexico, southward to northwestern Durango at 1,220 to 3,000 m ASL. The species is abundant on high-elevation grassy plains with bunch grasses within forests. It quickly reduces its numbers or disappears from areas where cattle graze. Slevin's Bunchgrass Lizard is usually associated with grass tufts used for cover, which cattle destroy.

Figure 58. Slevin's Bunchgrass Lizard. Photography JCM.

**REPRODUCTION.** As previously hypothesized, female Bunchgrass Lizards lay eggs instead of giving live birth. The eggs are laid when about one-third are developed and placed in a nest about six centimeters below the surface. Clutch size is 9 to 13 eggs laid in June to August. Incubation time is about 49 days.

# Tree Lizards, *Urosaurus*

Tree lizards and brush Lizards comprise the genus *Urosaurus*. They are native to North America, specifically the arid and semiarid regions of the western United States and Mexico, spending most of their time on trees, shrubs, or boulders.

Members of the genus *Urosaurus* can be distinguished from *Sceloporus* by the presence of a gular fold and granular lateral scales in *Urosaurus*. They can be distinguished from *Uta* by (sometimes only slightly) enlarged dorsal scales.

## Schott's Ornate Tree Lizard

*Urosaurus ornatus schottii* (Baird 1859) Figure 59.

**IDENTIFICATION.** In Arizona, the subspecies *U. o. schottii* covers most of the state and extends southward into western Mexico, and it is also present in western New Mexico it is perhaps one of the most commonly seen lizards due to its ability to colonize urban habitats and inhabit virtually all the biomes in the state except the Alpine Tundra. It ranges in elevation from sea level to 2,440 meters. Its small body (60 mm) and cryptic coloration protect it from predators. Often observed on trees, less often on boulders, and rarely on the ground. Both sexes defend territories against members of the same sex.

**DIET.** Almost all invertebrates it can capture are prey, including beetles, ants, crickets, grasshoppers, spiders, scorpions, and ticks. While this is primarily a diurnal ambush predator, it has been observed feeding at night. Sullivan and Rice (2021 reported American Kestrals in the Phoenix area were catching tree lizards to feed their young.

**REPRODUCTION**. Females lay clutches of up to 16 eggs in spring and summer. As many as seven clutches in one season have been reported. The hatchlings are present from May to September. Juveniles that hatch before September can reproduce the following spring. Hatchlings are 22 to 27 mm SVL.

**TAXONOMY**. Ten subspecies have been described, and at least three occur in Arizona: Schott's Tree Lizard, *Urosaurus ornatus schottii*, the Colorado River Tree Lizard, *Urosaurus ornatus symmetricus*, and the Northern Tree Lizard, *Urosaurus ornatus wrighti.*

**POLYMORPHISM.** Male Tree Lizards exhibit a dewlap color polymorphism associated with the male's mating behavior. Lattanzio et al. (2014) found blue morph males are aggressive and defend territories, yellow males are less aggressive and defend smaller territories, and orange males are typically nomadic. Females also show dewlap color polymorphism, but whether it is associated with mating behavior is unknown. Lattanzio et al. experimented with female mate choice and found that female preferences involved male dewlap color and the size of the ventral color patch. However, the female morphs responded to these signals differently and depended on the choice between the types of male morphs. Their experiment suggests females may distinguish male morphs using chemical signals alone. Yellow females based their male preferences on chemical and visual signals, while orange females preferred dominant males. They also found evidence for a chemical femoral-pore-secretion polymorphism in males. Genetic evidence suggests many cryptic species remain undiscovered in this taxon.

Side-blotch Lizards are small, omnivorous species that have a dark blotch of pigment

Figure 59. Schott's Ornate Tree Lizard. Photography JCM

behind their arm. Males tend to be more colorful than females. There are 12 species and subspecies in the genus.

## Western Side-Blotched Lizard

*Uta stansburiana elegans* Yarrow 1882 (SE) Figures 60-63.

**IDENTIFICATION.** Adults grow to 63 mm SVL, and the unbroken tail can be almost twice the body length. Hatchlings are 20 to 26 mm SVL.

The dorsum is brown, grey, yellow, or black, with dark scars, spots, and sometimes stripes. Smooth granular scales cover the dorsum, with larger scales on the head and limbs; a gular fold is present. Paired dark spots or wedges edged with white on the posterior edge are often present. The venter is white to grey and mostly uniform. The gular area is mottled dark and light. The common name is derived from a dark blue-black patch on the sides of the body behind the front limbs. This patch is sometimes indistinct or absent, but it will distinguish this species from all other Arizona lizards. Males have blue specks on the upper surfaces of the body, limbs, and tail. Distinct blue coloring on the belly is absent. However, there are three morphs in males. The throat is marked with blue, orange, or yellow. Males often have many blue speckles on the tail and the body's posterior. Females are drab in color. Their dorsum is blotched with brown and white stripes usually present, the side blotch is poorly defined.; and blue speckling on the throat is absent.

**DISTRIBUTION.** The Western Side-blotched Lizard ranges from central California, southern Nevada and nothern Arizona southward to the southern coast of Baja, and the west coast of Mexico to the Baja Peninsula and central Mexico. It occurs over most of Arizona but avoids high mountains and the Mogollon Rim. The elevation ranges from sea level to about 1,200 m. It uses all desert and grassland biomes. Sunny, warm days, even in winter. This usually diurnal lizard can sometimes be found basking on warm pavement at night. The function of this behavior is not well understood.

**DIET.** An insectivorous species. Prey includes ants, ant lion larvae, flies, mosquitoes, damselflies, dragonflies, beetles, bees, aphids, caterpillars, ticks, scorpions, and spiders. It is an ambush predator but will also forage for food. One estimate suggests that one of these lizards consumes 9,000 insects annually.

**REPRODUCTION.** Females are sexually mature at 37 mm SVL. Mating occurs mainly in the spring. Between March and August, females lay one to seven clutches of one to eight eggs. Hatchlings are exceptionally small and present from June to September. They can reproduce the following spring.

The species has an average life span of 4.8 years The clutch size may be as many as 12 eggs. Females lay their first clutch of eggs by April, and the incubation time is about 63 days. Females may lay as many as three clutches per season. Color variation in this species is dramatic, and there may be several color morphs in a population; each morph seems to have a slightly different reproductive strategy.

# The Skinks

The skink family, Scincidae, is diverse, with more than 1,613 species. Like the anguids, they have osteoderms, a layer of bony armor under their skin. They range in size from minute, with some species only 23 mm long, to the largest, which reaches almost a half meter. They are present on all land masses but usually avoid high latitudes and elevations. They

Figures 60-63. Western Side-blotch Lizard. Photography JCM

may be terrestrial, arboreal, fossorial, or semiaquatic. In the Australian deserts, skinks dominate the lizard fauna. However, in North American deserts, they are almost absent. Arizona has only five species of skinks, all in the genus *Plestiodon*, a clade with a Eurasian, North American, and Middle American distribution. Skinks of the genus *Plestiodon* are terrestrial throughout their range in eastern Eurasia, including Japan, China (including Taiwan), North and Middle America, and Bermuda. They use a variety of habitats, including deciduous forests, high plateaus, and subtropical islands. Their body forms range from the 'typical' stocky, robustly limbed lizard morphology to elongated, miniaturized, and reduced limb species. The genus contains many species formerly classified as members of the genus *Eumeces*.These secretive, agile animals have cylindrical bodies covered with smooth, shiny scales. They are distributed from East Asia to throughout North America, from southern Canada to Mexico, including oceanic islands such as Bermuda.

---

## Mountain Skink

*Plestiodon callicephalus* (Bocourt, 1879) (AS) Figure 64.

**IDENTIFICATION.** Adults grow to 76 mm SVL, with the tail about twice the body length. Hatchlings are 20 to 25 mm SVL. A pair of yellow stripes joined at the snout extends to the back of the head, making this species readily recognizable. The dorsum ranges from copper or bronze flecked with black to olive-grey. A distinct dark brown lateral stripe extends from the snout to the insertion of the rear legs. Adult males have bright red lips. Like most North American skinks, juveniles have bright blue tails; this coloration is retained in the adult Mountain Skink.

**DISTRIBUTION.** This Sky Island species uses the upper-elevation semi-desert grasslands and Madrean evergreen forest in the southernmost ranges of Pima and Cochise counties of Arizona and southwestern New Mexico. It ranges southward in the Sierra Madre to Nayarit and Jalisco, Mexico, and its elevation ranges from sea level to 2,000 m.

**DIET.** The diet has not been well studied, but they will take a variety of small arthropods. Females have been reported to tend their clutches of three to six eggs and, if disturbed, to move the eggs by picking them up in their mouths.

**ESCAPE BEHAVIOR.** The lizard quickly ejects its tail as a predator-escape mechanism. A long or bright blue tail can act as a predator-distraction device to draw a predator's attention to the tail by color or movement. The tail-loss device and the bony armor of these lizards combine to make them better able to deal with predators.

**REPRODUCTION.** The female Mountain Skink seems adapted to its high-altitude environment by retaining its eggs until the embryos are well-developed. Zweifel (1962) reported a clutch of three eggs that hatched after just nine days of incubation, whereas previous reports of incubation times had a minimum of 19 and 23 days. The nine-day incubation period may have been possible because the eggs were maintained at 87°F. Egg retention allows the female to manipulate the temperature of the developing embryos to maximize their growth.

**PARENTAL CARE.** There is additional evidence of extended post-hatching parental care. Cruz-Saén et al. (2011) observed the Mountain Skink in Jalisco, Mexico. Twice in July, they found female skinks under rocks with their eggs. In the first instance, the female had seven eggs, and in the second instance had three eggs. As the stones were turned, the females began transporting the eggs away from the nest. Perhaps they were seeking a more secure and stable thermal environment. The skinks were at 1,230 m ASL in an oak forest. Other names for this skink include *Eumeces callicephalus*.

Figure 64. Mountain Skink. Photography JCM

## Great Plains Skink

*Plestiodon obsoletus* Baird & Girard, 1852 (SE) Figure 65-66.

**IDENTIFICATION**. Adults grow to 142 mm SVL, with the tail 1.5 to two times the body length. Hatchlings are 36 to 41 mm SVL. A robust body, tiny legs, and a thick tail distinguish this skink from most, which is covered with smooth cycloid scales. It is tan to cream in color, with the edge of each scale bordered with black, producing a net-like pattern on the body and limbs. It may have a pink or violet wash. Occasionally, the black is reduced, and the dorsum appears to be flecked or spotted. Appendages may have a yellow or yellow-orange wash, and the ventrolateral areas may have orange or salmon markings. Hatchlings are jet-black with a blue tail, with white or yellow spots on the lips, and a uniform grey venter. This species is unique among the Arizona skinks in having obliquely arranged scales on the dorsum.The black coloration of the juvenile body and head may make the animal more difficult to observe, while the bright blue tail distracts the predator away from the body.

**DISTRIBUTION**. Ranges from northeastern Colorado, southern Nebraska, and southwestern Iowa through Arizona, New Mexico, Kansas, eastern Missouri, Oklahoma, Texas, and adjacent northern Mexico to northern Tamaulipas and Durango. The elevation range extends from near sea level to 2,650 m.

**HABITAT.** On the plains of eastern Colorado, these skinks are found in dense low vegetation, near rock outcroppings, or beneath trash and rocks (Maslin 1947). Moore & Rigney (1942) observed that the species frequently occurs in old buildings in Payne County, Oklahoma. The species is increasingly confined to rugged terrain in the southwestern part of its range. This holds true in the Santa Ritas where it inhabits mountain slopes and canyons. The Great Plains skink is rare or absent in the intervening stretches of desert plains. Bailey (1905) listed *P. obsoletus* as member of the Upper Sonoran Life zone. Little (1940) found it in the Roosevelt Reservoir area of central Arizona in the leaf litter of the chaparral woodland of the Upper Sonoran Life-zone but noted they were uncommon. Gloyd (1937) mentioned finding specimens beneath stones at 5400 ft. in Carr Canyon, Huachuca Mountains, and others in a stone pile on a sloping plain Quaintance (1935) recorded one found under a stone at the top of Rose Peak, 8787 ft. in Greenlee County, Arizona

A wide variety of mesic habitats are used. It prefers open habitats but is known to use forested habitats as well. It uses ground cover and rodent burrows for refugia. While it

Figure 65. An adult Great Plains Skink. Photography by Mason Salem.

likely evolved to inhabit prairies, it has been able to maintain populations in semi-desert environments where water is present.

**REPRODUCTION** Females produce a more oversized clutch of eggs than other Arizona skinks, likely related to their size. Hatchlings, present in July and August, are black and grow rapidly in their first year, with growth slowing as they age (Bateman and Chung-MacCoubrey, 2011 Sexual maturity may not be reached until three years of age. Carron and Swan (2007) found females mature at about 3 years of age, they produce a single clutch of 7-24 eggs annually, with larger females producing more eggs. Hatchlings have a body length of 33-43 mm and are present in July. Males mature after their third hibernation and develop muscular enlargements on each side of the head during the breeding season.

**DIET.** Eastern populations have been studied for diet composition. One study reported grashoppers and caterpillars making up 99% of their prey. A second study found that much of the diet is comprised grasshoppers, spiders, and beetles.

**ACTIVITY**. While other Arizona skinks are diurnal, this species is reported to be partially nocturnal. Active individuals have been observed in late afternoon by this author, but not after dark.

# Whiptails

Teiids can be distinguished from other lizards by the following characteristics: large rectangular scales form distinct transverse rows on the belly, with small granular scales covering the dorsum; head scales that are separate from the skull bones, and teeth that are solid at the base and "glued" to the jaw bones.

Additionally, all teiids have a forked, snake-like tongue and well-developed limbs. Teiids are terrestrial, but a few are semi-aquatic. They are diurnal and primarily carnivorous or insectivorous, but they include some plants in their diet. Most teiids forage actively within their ideal temperature range, quickly skirting between cover objects. They are ovipa-

Figure 66. A hatchling Great Plains Skink. Photography by Susan Myers.

rous, and some species lay huge clutches.

Only one genus occurs in Arizona, *Aspidoscelis*. About 11 species of whiptail lizards are in the state, and one appears to have been introduced. The whiptails have the upper surface of their body covered with tiny granular scales, and the belly is covered with large plate-like scales in eight longitudinal rows. They are fast-moving when escaping a potential predator but walk with a jerky gait, as they occasionally pause, raise one foot off the ground, and shake it. Their head is in constant motion, moving from side to side, and they stick their head under leaves or other debris piles looking for food. Arthropods are a major portion of the diet, a few eat fruits.

Some Arizona whiptails are all-female species that reproduce by parthenogenesis. These parthenogens evolve from sexual species (males and females present in the population) hybridizing. Many whiptails are difficut to identify. Most Arizona species have a striped pattern that may or may not have spots. The number of stripes and the position of the spots are often crucial in identification. Whiptails are active diurnal foragers. Female whiptails lay clutches of six eggs or less. There is evidence that some high-elevation populations may have live births or very short incubation times for their eggs. The North American whiptails were formerly placed in the genus *Cnemidophorus* but have all been moved to the genus *Aspidoscelis*.

---

## Sonoran Whiptail

*Aspidoscelis sonorae* (Lowe & Wright, 1964) (SE) Figures 67-68.

The populations previously called *A. flagellicauda* are now placed in this species.

**IDENTIFICATION.** Adults reach 91 mm SVL, with the tail about twice as long as the body. This is a six-striped, spotted whiptail. Pale spots are found between the stripes. Mottling on the upper thigh and forelegs may be present, or not. The tail is orange-tan, brown-tan, or blue-tan. The venter is white to cream. Spots are reduced or absent in juveniles, while limb markings are more prominent. It is most easily confused with the larger Canyon Spotted Whiptail and the Chihuahuan Spotted Whiptail, which both present spots on the upper surfaces of the rear legs.

**DISTRIBUTION.** The Sonoran Spotted Whiptail is a triploid, all-female species ranging from southeastern Arizona and southwestern New Mexico southward into Sonora at 610 to 2,100 m. This is a desert grassland species that ranges into pinion-oak woodland. In drier habitats, it stays near riparian corridors. It is active from April through October and maintains a body temperature of about 40°C. It is common in suburban communities built in semi-desert grasslands.

**DIET.** Termites compose most of its diet, with spiders, beetles, ants, and grasshoppers on the menu. It will scavenge dead insects as well as capture live ones. Food is often dug up or captured by moving leaves or other debris on the ground. Termites have been reported as an essential part of its diet. However, it will also chase prey. One individual was observed preying on a small Desert Iguana. Active lizards have body temperatures of 39°C, but once they reach 42°C, activity ceases, and they head for shade or a burrow.

Figure 67-68. Sonoran Whiptail. Photography JCM

**REPRODUCTION.** Clutch sizes range from 1 to 7, and the average clutch size is about four eggs. Hatchlings first appear in July. Its parental species seem to be *A. stictogrammus*, *A. costata*, and a member of the *A. inornatus* complex.

**REPRODUCTION.** In Arizona, Tiger Whiptails reach high densities; 36 individuals per hectare have been reported. Females reach sexual maturity at 60 mm, males at 70 mm. Clutch sizes vary from one to five; some females produce two clutches per year.

**ACTIVITY.** They are active in the morning, and their seasonal activity usually extends from April to August. It occurs at elevations that range from sea level to 2,000 m.

Sonoran Tiger Whiptails are hyperactive and can often be identified at a distance just by how they move: the body is up on all four limbs, and the tail is straight out as they rush away from the observer.

**PREDATORS.** The primary predators of this lizard are the Leopard Lizard, the Coachwhip, and the Sidewinder, all sit and wait to ambush the whiptail as it hunts (Anderson, 1993).

## Sonoran Tiger Whiptail

*Aspidoscelis tigris punctilinealis* (Dickerson 1919) (SE) Figure 69.

Other common names include the Western Whiptail.

**IDENTIFICATION.** Adults reach 120 mm SVL, with a tail two or three times the body length. Males tend to be more massively built than females, even when they have the same body length. Walker et al. (2021) suggest treating *A. tigris* as a polytypic or several gonochoristic species.

This species group ranges from southwest New Mexico, across southern Arizona, and into southeastern California

This is a striped and spotted whiptail with a pattern that fades with age. Three of the five recognized subspecies are present in Arizona. The subspecies are based chiefly on coloration, which is quite variable. In the focus area the Sonoran Tiger Whiptail, *Aspidoscelis tigris punctilinealis,* is present.

**DISTRIBUTION and HABITAT.** This is a Sonoran-Chihuahuan desert species mostly restricted to southeast Arizona and Sonora, Mexico. It is abundant in semi-desert grasslands, including ones that have been urbanized. Its dark throat, dark chest, four stripes, and spots make it readily recognizable.

**DIET.** The foraging behavior requires it to use vision and chemoreception to locate food.

---

## Desert Grassland Whiptail

*Aspidoscelis uniparens* (Wright & Lowe, 1965) (SE) Figure 70.

**IDENTIFICATION.** Adults grow to about 86 mm SVL, with a tail twice as long as the body. It has six pale stripes without spots on the darker pigmented fields between the stripes. A seventh, indistinct stripe may be on the anterior and posterior body. The tails of adults are tan to brown, and the dorsal stripes may extend onto the tail at its base. The distal portion of the tail has an olive-green or blue-green tint. Hatchlings and juveniles have a more vivid pattern with a blue to blue-green tail. Ventral surfaces are white, but some may have a blu-

Figure 69. Sonoran Tiger Whiptail. Photography JCM

ish wash on the head and anterior body. The problem with this description and the extant field guides is that they describe the distil portion of the tail as being blue or blue-green. Most of the specimens from the focus area have brown tails, as in Figure 58. In fact look at the photos in iNaturalist and most of them have brown tails. Given that the type locality is in Cochise County relatively close to the focus area it raises questions about the idenity of the brown tail-morphs *uniparens.*

**DISTRIBUTION and HABITAT.** This is an all-female species of grasslands and semi-desert grasslands. It occurs with other whiptail species in arroyos and riparian corridors at elevations below 1,500 m. It is distributed southward from southeastern Arizona and central New Mexico into Sonora and follows the Rio Grande drainage into Texas. It is most abundant in grasslands that have been degraded with shrubs. It may reach densities of 78 to 103 individuals per hectare. Seasonal activity is from March to November. Its daily activity pattern is bimodal, with activity in the morning until about 11:30 a.m. and again between 5:30 p.m. and 7:30 p.m.

**REPRODUCTION**. Captive specimens behave as males and have pseudocopulatory behavior; the same individual may act as a female or a male. Female-like behavior seems to be associated with individuals ready to lay eggs. False-copulatory behavior may trigger ovulation, but this has not been studied in wild populations. Individuals become reproductive at about 60 mm in body length. Clutches of one to four eggs (an average of about three) are laid from late spring to mid-summer. They reproduce in their first year, and few survive their second year. Thus, this seems to be an annual species. Roadrunners, owls, shrikes, and leopard lizards are important predators.

**NATURAL HISTORY**. Mitchell (1979) studied four species of whiptail lizards, two bisexual (*Aspidoscelis inornatus* and *A. tigris*) and two unisexual (*A. sonorae* and *A. uniparens*), for ecological differences in the semi-desert grassland of southeastern Arizona. Density estimates indicate the decreasing order of abundance was *A. uniparens*, *A. inornatus*, *A. tigris,* and *A. sonorae.* Daily activity was generally bimodal, and active lizards were usually encountered away from vegetation. The use of shade patches differed among species. Termites, beetles, orthopterans, and insect larvae dominated their diets, with other kinds of prey taken in smaller quantities. Ecological differences between the bisexual and unisexual species were found to be minimal. Niche overlap in the order of importance in separating these four whiptail niches is macrohabitat, microhabitat (this is redundant), time, and food. These four whiptails are ecologically similar, and they form a guild. Wright and Lowe (1968) hypothesized that the asexual species evolved in environments that were disturbed or in transition, similar to the environments that produce plants referred to as "weeds."

## Canyon Spotted Whiptail

***Aspidoscelis stictogrammus*** (Burger 1950) Figure 71 -72.

**IDENTIFICATION.** Canyon Spotted Whiptails differ from other whiptails because of their enormous size and many spots. In our area, no other whiptail reaches a size of 120 mm SVL or more, nor does any have more than 90 scales surrounding its midbody. Even though it is diagnostic, the latter trait is useless for casual lizard observations. Juveniles have stripes, and there may or may not be light patches in between the stripes. This results in young individuals being frequently mistaken for other species. However, hatchlings and young specimens of the Canyon Spotted Whiptail can be distinguished from other species in the area by its orange to orange-red tail as a hatchling and juvenile. Large, mature So noran Whiptails (*Aspidoscelis sonorae*) always retain stripes, at least on the neck; light spots

Figure 70. Desert Grassland Whiptail. Photography JCM

may infrequently take over the That species never has a tail as orange or orange-red as the Canyon Spotted Whiptail; instead, its tail is tan, orange-tan, or brown, occasionally with an olive tint at the end. The whiptail previously known as *A. xanthonota* are now part of this species.

This large whiptail is challenging to approach or watch. Most are found between April and September, but they have been seen as early February to as late as mid-October. In Pima County, Goldberg (1987) discovered they come out of hibernation in April and early May; and after early September, the adults are seldom observed. These lizards are easy to miss and rarely seen, despite their large size.

**DIET**. The Canyon Spotted Whiptail prey on a wide range of invertebrates and may eat tiny lizards. Termites and ants were the most important food items at two sites in northeastern Sonora. Caterpillars, beetles and their larvae were also significant prey. One stomach contained plant material.

**REPRODUCTION**. Mating occurs in April and May, in June and July, they lay clutches of three to ten eggs. In years with plentiful food, two clutches are possible Sexual maturity is reached in males >89 SVL and in females >90 mm SVL. The appearance of hatchlings, is in early August. They are about 33 mm SVL (Degenhardt et al., 1996).

Figure 71-72. Canyon Spotted Whiptail.
Top Photography JCM. Bottom Photography Michael Andersen

# Serpentes

Snakes are carnivorous, elongated reptiles found on every continent except Antarctica. More than 4145 species of snake are known. Snakes evolved from lizards and together they make up the order Squamata. It is estimated that about 10% of snake species globally are threatened with extinction. However, snakes are secretive and many species are known from a very small number of specimens. Thus, 10% is likely a low estimate for threatened species. They are limbless, but some families have retained remnants of limbs that may be internal (Threadsnakes) or mostly internal with an external element (Boas).

# Threadsnakes

The Leptotyphlopidae (slender blind snakes or threadsnakes) are a family of snakes found in North America, South America, Africa, and Asia. All are fossorial and adapted to burrowing, feeding on ants and termites. Two subfamilies are recognized, one in Africa and one in the western hemisphere.

Small snakes, threadsnakes rarely exceed 12 inches in length. The cranium and upper jaws are immobile, and the upper jaw lacks teeth. The lower jaw consists of a much elongated quadrate bone, a tiny compound bone, and a larger dentary bone. The body is cylindrical, with a blunt head and a short tail. The scales are highly polished. The pheromones they produce may protect them from attacks by termites. These snakes are oviparous, and some species show parental care for their eggs.

Screech owls capture threadsnakes and bring them to their nest to feed their chicks. But on occasion, the snakes escape the birds and take up residence in the nests, feeding on soft-bodied insects that parasitize the nestling birds. Gehlbach and Baldridge (1987) suggest nestling owls that lived with threadsnakes grew faster and had a lower mortality rate than broods that were not living with threadsnakes.

In this section I am following Flores-Villela et al. (2022) who recognize three species of the genus *Rena* in, or close to, the focus area. They wrote, "The central Texas *R. dulcis* and southeastern Arizona *R. dissecta* show only a 5% sequence difference, a very small distance for samples coming from populations more than 630 km apart, regarded here as within species variation. Between *R. dissecta* or *R. dulcis* and *R. klauberi* there is an 11% sequence divergence, attributed to a species level difference. Species from the *R. dulcis* group differed from those within the *R. humilis* group by between 14-15%, genetic distance.",

---

## Trans-Pecos Threadsnake

***Rena segrega*** (Klauber 1939) (?)

**IDENTIFICATION.** *Rena segrega* has 253-287 middorsal scales; 14 rows of scales around midbody; it has 10 rows of scales around the base of the tail; it has no supraocular scales; parietal and posterior labials are in contact; seven pigmented rows of dorsal scales; maximum size 319 mm.

**DISTRIBUTION.** The species ranges from southeastern Arizona, New Mexico, and the Trans-Pecos area of Texas southward into Mexico to San Luis Potosí, Coahuila, Nuevo León, Durango, and Chihuahua.

## Texas Threadsnake

***Rena dulcis*** Baird and Girard 1853 (?)

The presence of this snake in the focus area is problematic.

**IDENTIFICATION.** *Rena dulcis* has 192-257 middorsal scales; 14 rows of scales around midbody; it has 10 scale rows around the tail; it can have one or no supraocular scales; parietal and posterior labials in contact; seven (5-8) pigmented rows of dorsal scales; maximum size 272 mm.

**DISTRIBUTION**. On the eastern edge of it range it extends from southern Kansas to Veracruz, Mexico and on the western edge of its range from southeast Arizona southward to Veracruz, Mexico.

## Southwestern Threadsnake

*Rena humilis humilis* (Klauber, 1939) (SE) Figure 73-74.

Note that this snake has also been called the Desert Blind Snake, Desert Threadsnake. Nine subspecies have been recognized.

**IDENTIFICATION**. Adults can reach 339 mm in total length. Hatchlings are about 90 mm in total length. The dorsal coloration is pink to red-brown to purple. The head is the same width as the body. An ocular scale covers the eyes and eyes appear only as small dark spots under the scale. The tail is blunt and has a downward curved spine. Scales are in 14

Figure 73-74. Southwestern Threadsnakes. Photography JCM

rows and are uniform in size around the body; they are glossy and stiff. One scale between the supraoculars distinguishes it from most other species in the genus with three scales in that position.

It has 12 scales around the base of the tail, 257-283 to 312 dorsals, and 15 to 21 subcaudals. The Southwestern Threadsnake, *Rena humilis humilis,* is found in northwestern and south-central Arizona and has 12 scales around the base of the tail, 257 to 283 dorsals, and 15 to 21 subcaudals.

**DISTRIBUTION.** The species ranges from southern California, southern Nevada, southern Arizona, southern New Mexico, and southern Texas, into Baja California, as far south as Nayarit, Mexico. The northern portion of this range has the subspecies *R. h. humilis.*

**HABITAT.** The Southwestern Threadsnake is known to use desert scrub to interior chaparral biomes. This species can make its own burrow, but it will also use burrows made by other animals.

**DIET**. Ants and termites made up 54% of stomach contents and soft bodied arthropods composed the remainder of the diet (Punzo 1974).

**REPRODUCTION**. Females that reach 245 mm in total length are sexually mature.Clutch sizes are 2-8 with larger females having more eggs. Incubation requires 94-105 days. Hatchlings are 65-127 mm in total length. May lay eggs in a colony with females in attendace

**FIELD NOTES** Found crossing roads and will behave very erratically, thrashing from side to side while releasing anal musk.

In some habitats and under some conditions, these snakes are exceptionally abundant. Surface activity is influenced by ant and termite presence, moist soil, and moonlight. Threadsnakes will follow the chemical trails laid down by ants or termites to locate their nests. Exactly how they avoid the bites of soldier ants and termites is unclear. Some have suggested that they raise the scale edges; others suggest that they produce a chemical that mimics the insects' odor.

## The Superfamily Colubroidea

The following species are members of the superfamily Colubroidea. They are organized by subclades, groups that share a relatively recent ancestor. They include the vipers and elapids, both with front fangs and venom glands. They inject prey with venom under high pressure. Other sub-clades found within this superfamily are often considered non-venomous, yet many have fangs on the back of the maxillary bone and venom glands (these are rear-fang snakes with an open groove on the fang that conducts venom into the prey under low pressure. These snakes are referred to as rear-fanged snakes. To inject venom into their prey, they often chew while holding the prey in their mouth. In the United States, most rear-fanged colubrid snakes are considered harmless to humans, but some bites may result in a mild reaction.

## Gartersnakes

The Natricidae (or natrids) is a worldwide family of snakes, sometimes referred to as a subfamily (Natricinae) of the colubrids. The clade comprises about 275 species in 36 genera. Members include many very common snake species, such as the European grass snakes, and the North American water snakes and garter snakes. Some Old World members of the subfamily are known as keelbacks because their dorsal scales exhibit strong keeling (as do many of the North American species).

Gartersnakes and their relatives, often called water snakes, have a Holarctic distribution with about 29 genera and 210 species across Asia, Europe, North Africa, sub-Saharan West Africa, and North and Central America. Many snakes are associated with aquatic habitats and feed on fish and frogs. Some have adapted to xeric conditions and may feed on invertebrates or mammals. Molecular studies suggest the natricines evolved from a colubroid ancestor in the Eocene or Oligocene.

Gartersnakes of the genus *Thamnophis* are North American members of the Natricidae. Natrids occur in Eurasia, North America, Central America, and Africa. As a group, they seem more tolerant of low temperatures than other snakes. The Western Hemisphere clade Thamnophiine contains about 60 species from Canada to Costa Rica. They tend to feed on soft-bodied prey, but a few have adapted to feeding on crayfish. All North American members share an ancestor that originated in Asia and colonized North America, probably across the Bering Land Bridge sometime in the Miocene, 14–11 MYA.

Gartersnakes are viviparous, and females give birth to live young, nourishing them with a combination of yolk and nutrients supplied from the female's bloodstream through a placenta. Arizona has five species of gartersnakes. Four of Arizona's gartersnakes have three longitudinal yellow stripes and are easy to confuse; some may occur together. By determining which lateral scale row has the yellow stripe on the anterior of the body, it is possible to distinguish them. The Black-Necked Gartersnake has a yellow stripe on rows two and three; the Mexican Gartersnake has a lateral stripe on scale rows three and four, and the Checkered Gartersnake has a stripe restricted to row three. Gartersnakes and most other thamnophiines have females that are larger than males. The only non-gartersnake natricine present in Arizona is the recently introduced Banded Watersnake. It occurs in southwest Arizona.

---

## Black-Necked Gartersnake

*Thamnophis cyrtopsis* (Kennicott 1860) (AS). Figures 75-76.

**IDENTIFICATION.** Adults grow to 1,070 mm, with the tail 22 to 27% of the total length. Neonates are 150 to 280 mm in total length. The head is wider than the forebody. The dorsum is a dark olive-grey with an orange-yellow vertebral stripe and a lateral cream or white stripe on scale rows 2 to 3. There are 19 rows of keeled scales at mid-body. Two lateral black blotches are immediately behind the head, giving this snake its common name. Ventral surfaces are uniform grey or cream. Black pigment is present on the upper labial seams. The stripe on rows 2-3 and the pair of black blotches distinguish this snake from other Arizona gartersnakes. The Mexican Gartersnake and the Checkered Gartersnake have 19 scale rows, but the yellow stripe is on rows 3-4 and scale row three, respectively. The Terrestrial Gartersnake and the Narrow-Headed Gartersnake have 21 mid-body scale rows.

**DISTRIBUTION.** The distribution ranges from Utah and Colorado southward to Guatemala. While they follow streams and stay close to water, they will wander overland a kilometer or more from a water source in deserts or coniferous forests.

Like other gartersnakes, they have venom glands and venom, but they are not considered dangerous to humans. The venom is used to help them subdue prey, mostly frogs (particularly the Red-Spotted Toad and Chiricahua Leopard Frog), although neonates and juveniles will feed on worms and small fish. This snake will coil in a stream behind rocks, waiting to ambush prey. In Arizona, it is active from March to October and has an average body temperature of 26 to 27°C. Its activity is primarily diurnal (McCall et al. 2017).

Sexual maturity is reached in males at about 34 cm and in females at about 41 cm. Court-

Figure 75-76. Blacknecked Garter Snake. Top photo JCM. Bottom Photo Alan Schmierer..

ship and mating occur in the spring and fall. The young are born from June to August, and the average litter size is about 10, ranging from 3 to 24 young per litter.

## Mexican Gartersnake

*Thamnophis eques megalops* (Kennicott, 1860). (CI) Figure 77

**IDENTIFICATION.** Adults grow to 1,120 mm in total length, with the tail 22 to 27% of the total length. Neonates are 225 to 258 mm in total length. The head is wider than the forebody, and keeled dorsal scales are found in 19 rows at mid-body. The dorsal coloration is red brown to olive-brown with a vertebral cream-colored stripe. There is a lateral stripe on scale rows 3 to 4. Between the dorsal and lateral stripes are small dark dashes in two rows. The venter is pale yellow or cream. Black pigment outlines the seams of the labial scales. These traits distinguish the Mexican Gartersnake from other Arizona gartersnakes.

**DISTRIBUTION.** In southern Arizona and New Mexico southward across the Mexican Plateau and Highlands. In the past, it was found at low- to mid-elevations (530 to 1,875 m) throughout the Gila River Basin and the headwaters of the Rio Yaqui in the San Bernardino National Wildlife Refuge. It was once common in the Tucson Basin. It has been extirpated from the area due to the alteration and destruction of its aquatic habitats and the introduc-

Figure 77. The Mexican Gartersnake (also called the Brown Gartersnake.)
Photography Larry L.C Jones.

tion of bullfrogs.

**HABITAT.** The Mexican Gartersnake is highly aquatic and rarely moves far from permanent water with dense aquatic vegetation between 530 and 1,875 m elevations. The diet includes invertebrates, fish, amphibians, reptiles, and mammals. It is also preys on largemouth bass. These game fishes are invasive in Arizona and are a serious threat to this vulnerable species. There is some anecdotal evidence that the Mexican Gartersnake may exclude the Black-necked Gartersnake from a habitat that contains native fish species and may have some structural differences (wider, shallower streams) (O'Donnell et al. 2016).

**CONSERVATION STATUS**. Surveys conducted during the 1980s and 2000 suggest that this trend continues. On 8 July 2014, the Brown Gartersnake was listed as a threatened species under the U.S. Endangered Species Act. The population in Santa Cruz County, previously believed extirpated, was recently found to survive at low population densities. As many as 24 of the 29 known Brown Gartersnake localities in the United States (83%) are thought to be non-viable; if the snake is present, it may exist only at low population densities that could be threatened with extirpation or may already be extirpated. Bullfrogs, crayfish, and introduced game fish seem to be the culprits responsible for its decline. The introduced fish have eaten and outcompeted their way through the native fish, the major prey base for this species. Bullfrogs and crayfish are additional predators on the Mexican Gartersnake.

Only five populations in the United States are considered viable. In 2008, the total number of stream miles in the United States that historically supported the Brown Gartersnake was evaluated. Most of these miles are now permanently dry, except for temporary flows from heavy precipitation. The conclusion was that the subspecies has been extirpated from or occurs at low densities in as much as 90 percent of its historical range in the United States.

**REPRODUCTION**. Females mature at about 530 mm in total length and three years of age. Males probably mature at two years of age. Females likely reproduce every other year. Clutch sizes range from 4 to 38, with an average of about 14. The Mexican Gartersnake is the sister to the Checkered Gartersnake, *Thamnophis marcianus*. There is no open season on this snake. Endangered Species Act status is pending.

## Checkered Gartersnake

*Thamnophis marcianus* (Baird & Girard, 1853) (AS) Figure 78.

Adults grow to 1,080 mm total length, with the tail 19 to 26% of the total length. Neonates are 123 to 279 mm in total length. The head is wider than the forebody. The dorsum has 19 rows of keeled scales with apical pits. The dorsum has a yellow vertebral stripe, and a lateral stripe on scale row three is usually present (some specimens may lack this lateral stripe). Rows of alternating black blotches are arranged in a checkered pattern laterally. A dark blotch is located immediately behind the head, with a light area between the dark blotch and the rictus. The ventral surface is a uniform, pale cream or grey, which may have some faded dark pigment. These traits distinguish the Checkered Gartersnake from other

Figure 78. Checkered Garter Snake. Photography JCM

Arizona gartersnakes.

The subspecies present in Arizona is Marcy's Checkered Gartersnake, *T. m. marcianus*. Its distribution ranges from Kansas southward to Veracruz, Mexico, and westward to southern California and Baja. In Arizona, it is present in the southern third of the state. This is a lowland species that can reach altitudes of 1,700 m above sea level (ASL). It is often found in association with watercourses, roadside ditches, ponds, and irrigated agricultural land. It will travel overland to find water. It is associated with the Sonoran and Chihuahuan desert scrub, semi-desert grasslands, and plains grasslands. Its diet is diverse: it preys upon worms, molluscs, crustaceans, fish, amphibians, and reptiles.

The ecology of Marcy's Checkered Gartersnake was studied at Roper Lake State Park in Graham County, Arizona (Seigel et al., 2000). The park is in a desert grassland and the researchers examined the population, using a man-made pond. The snakes were active from March to October and could be active at any time of day. Snakes were always near the water, and only four specimens were found at distances greater than 10 m away from the water. Mating occurred in late March, and females gave birth between late May and early June. Litter size averaged 15 young. Like other gartersnakes, males scrambled to mate with females and form mating balls. The smallest gravid or newly spent female was 515 mm SVL, and the most petite male found in the mating ball was 478 mm. This gartersnake seems to be doing relatively well, compared to the Brown Gartersnake, probably because it is more terrestrial and therefore avoids bullfrogs.

This species is listed as Threatened by the US Fish and Wildlife Service, and it cannot be collected without special permits.

# Dipsadid Snakes

The more than 750 species of the Western Hemisphere Snakes, the Dipsadidae (or Dipsadidae) are mostly restricted to the Neotropics, with just a few species entering the United States. One genus, *Thermophis,* inhabits the Tibetan Plateau and suggests the family's ancestry is in Asia. At least five species occur in Arizona, and one remains undescribed.

# Ring-Neck Snake

*Diadophis punctatus,* commonly known as the Ring-necked Snake, is a small to medium-sized snake. It is harmless to humans but is venomous. It is considered to be a single polytypic species with 14 recognized subspecies. It occurs as far north as Quebec, Canada, southward throughout much of the United States, and extends southward to Central Mexico ( San Luis Potosí). Ring-necked snakes are fossorial and secretive. They are nocturnal and rarely seen during the day but may be found under cover objects. Its unique defensive posture involves curling the tail tightly into a coil and partially rolling onto its back, exposing its bright red-orange underside and ventral surface. IO'Donnell et al. (2007) provided experimental evidence for venom in this species.

Fontanella et al. (2008) proposed the species contained 14 lineages. Many seem restricted to specific habitats, such as grasslands, floodplains, and montane settings. Diversification in the USA started in the late Miocene and lasted until the mid-Pleistocene, indicating that these lineages existed before the last glacial maximum.

---

## Regal Ring-Necked Snake

*Diadophis punctatus regalis* Baird and Girard 1853 (LC) Figures 79-80.

**IDENTIFICATION**. These are highly distinctive snakes with a brightly colored collar and ventral surface. The Regal Ring-neck Snake has seventeen scale rows at mid-body. There is usually a yellow or orange collar just behind the head. The dorsum is usually a uniform brown-grey or olive-grey, with a yellow or orange ventral surface and scattered darker pigment. The ventral, posterior body, and tail are bright red-orange with darker pigment. The maximum size is 940 mm in total length.

**DISTRIBUTION**. *Diadophis p. regalis* ranges from southeast Idaho south to Sonoran, Mexico, and is present in Utah, Nevada, Arizona, New Mexico, west Texas, and south into Mexico (Sonora, Chihuahua, and Coahuila)

In Arizona *Diadophis* uses montane habitats, deserts, grasslands, and forests to at least 2,200 m. It seems to be most abundant in areas with surface soil moisture; and it can be active at any time of day.

**DIET.** Regal Ring-necked Snakes' diet includes other snakes and lizards. They have rear

Figure 79-80. Regal Ringneck Snake
Photography JCM

fangs and venom glands. The venom is fatal to neonate specimens of several other snake species, and high levels of phospholipase A2 were detected in its venom (Hill and Mackessy, 2000). The venom is probably not a serious problem for humans.

**DEFENCE.** Defensive behavior includes hiding its head and coiling and raising the tail to expose the bright coloration. They will also release a potent anal musk.

**TAXONOMY NOTE.** Fontanella et al. (2021) used species delimitation methods and failed to recover groups consistent with designated subspecies. Instead, they found well-supported, mostly geographically isolated lineages. Divergence time estimates suggest that the divergences were driven by historical isolation associated with Pleistocene climate shifts. They also found a correlation between increased morphological differentiation and time since divergence, and a more remarkable niche similarity between the more recently diverged eastern California and western California groups. They propose that the morphological similarities are due to a combination of morphological conservatism and evolutionary stasis. The study provides the foundation necessary to reassess the biodiversity and conservation status of ringneck snakes. It offers an essential step in unveiling the diversity within the western portion of the genus's range.

## Night Snakes

*Hypsiglena* is a rear-fanged genus of small, venomous dipsadid snakes; rarely do they exceed 40 cm. The genus holds nine species, three polytypic and distributed in western North America, ranging from southwest Canada through western USA, into Mexico on both sides of the Mexican Plateau. Members of the genus are referred to as Night Snakes. Night Snakes prefer semiarid desert regions with rocky and sandy soils and have a diverse behavioral repertoire. Night snakes are venomous, with rear fangs and a venom gland. Venom kills prey, but it is not usually considered harmful to humans. Nightsnakes should be handled carefully.

---

### Hooded Nightsnake

*Hypsiglena* sp. Figure 81.

**IDENTIFICATION**. Adults likely grow to 650 mm in total length, with a tail that is 17 to 21% of the total length. Hatchlings are about 150 mm in total length. Males are smaller than females. The head is distinctly wider than the forebody and is relatively flat. The dorsum is grey or tan with two rows of dark, grey-brown blotches on the vertebral line. Note that some of these may be fused into one blotch. Laterally, there are two more rows of much smaller spots. The neck is marked with a broad collar composed of a single blotch or a blotch divided into two or three parts, rounded on the back edge. A mid-dorsal stripe extends anteriorly from the front edge of the collar onto the crown. A dark grey-brown bar extends back from each eye and tapers to a point where it meets the collar. The belly is a plain, pale grey. The head is depressed. The pupils are vertically elliptical, and the scales are smooth.

**DISTRIBUTION.** The Hooded Nightsnake is relatively common in habitats ranging from

Figure 81. The Hooded Night Snake. Photography JCM

deserts to montane forests between sea level and 2,200 m ASL. During the day, it seeks cover in rock crevices, under surface objects, and in mammal burrows. The species occurs throughout the focus area. It is also present in Cochise, Santa Cruz, and Pima counties and appears to be in Pinal County near Marana, the Avira Valley, and Iron Wood National Monument. It goes southward into Sonora, Mexico.

**HABITAT**. Sonoran Desert Scrub, grasslands, forests, and the cold Petran Montane Conifer Forest, are among the habitats used. The terrain ranges from flat, open deserts to steep, rocky, forested slopes, it can be found in an equally diverse range of terrain types. It appears most prevalent in desert scrub and semidesert grassland habitats. Its diet and reproduction have not been studied. As the common name suggests, these snakes are nocturnal and likely spend the day in subterranean refugia, such as mammal burrows and rock crevices.

**DIET.** The diet includes other snakes, lizards, and frogs.

**PREDATORS**. Many mammals, birds, and reptiles are predators, as are Black Widow Spiders, which are known to prey on hatchlings.

**REPRODUCTION**. Females lay eggs in June and July, with hatchlings emerging in August and September. This species' eggs are laid in crevices in rocky outcrops, abandoned mammal burrows, and human-modified environments.

Note: This is an undescribed species following Mulcahy (2008).

## Colubridae

Colubridae or Colubrids is the most prominent family of snakes, with about 250 genera—the earliest fossil species of the family date to the Late Eocene. Colubrid snakes are found on every continent except Antarctica.

Colubrids have limbless bodies (no remnant hind legs), a reduced or absent left lung, well-developed oviducts, premaxillaries that lack teeth, maxillary bones oriented longitudinally with solid teeth, but some (rear fangs) have open grooves on the anterior surface. The mandible without a coronoid bone, dentary that has teeth, only a left carotid artery, intracostal arteries arising from the dorsal aorta every few trunk segments, and no infrared receptors occurring in pits or in surface indentations. Most species lay eggs as opposed to giving live birth. Colubrids have adapted to a wide range of life zones.

Some clades previously recognized as subfamilies have been considered families of the large superfamily Colubroidea by others.

## Leaf-Nosed Snakes

Two species of leaf-nosed snakes inhabit Arizona. They are small, less than 51 cm in maximum length, and very secretive. Most of the year is spent in underground refuges, and they likely surface to mate and hunt in April to June. The two species, described in the late 19$^{th}$ century, remained poorly known until 1922, when Lawrence Klauber discovered they could be found by driving asphalt roads at night. Both species share a vertical, elliptical pupil, a characteristic found in very few North American snakes. The leaf-nosed snakes are sisters to the lyresnakes (*Trimorphodon*). Both leaf-nose species occur in our area.

## Saddled Leaf-Nosed Snake

*Phyllorhynchus browni* Stejneger, 1890 Figures 82-83.

**IDENTIFICATION.** Adults reach 508 mm in total length, with the tail 7 to 13% of the total length. The head is slightly wider than the forebody. The rostral scale is thick and turned up and back over the snout. A dark mask comes over the crown and through the eyes to the mouth. The dorsum is grey, tan, or cream with 12 to 16 transverse saddles. The dorsal scales are smooth and in 21 rows. The eyes are large, and the pupils are vertically elliptical. The pattern distinguishes this species from the Spotted Leaf-nosed Snake and other snakes with modified rostral scales.

Figures 82-83. Saddle Leaf-nosed Snake. Photography JCM

**DISTRIBUTION.** It ranges from southcentral Arizona southward to Sinaloa, Mexico. In Arizona, it occurs in Maricopa, Pinal, Pima, and southeastern Yuma counties. The Saddled Leaf-Nosed Snake is associated with sandy, gravel soils with paloverde, mesquite, saltbush, and Saguaro.

**DIET.** Like its relative, the Spotted Leaf-Nosed Snake, is nocturnal, shelters in rodent burrows, and most likely feeds exclusively on lizard eggs.(Mitchell and Rosen 2020).

**REPRODUCTION.** Goldberg (1996) found the smallest mature female was 229 mm, and the smallest mature male was 293 mm. Mating occurs in the spring, and eggs are probably laid in August. Not all females reproduce every year. Clutches of 4 to 5 eggs have been reported.

**ACTIVITY.** The Saddled Leaf-Nosed Snake is active at air temperatures between 22 °C and 33°C and body temperatures of 31°C have been reported.

**DEFENSE.** When molested, it vertically flattens its neck and hisses. If the harassment continues, it tries to hide its head.

---

## Spotted Leaf-Nosed Snake

*Phyllorhynchus decurtatus* (Cope, 1869). Figure 84-85.

**IDENTIFICATION.** Adults reach 510 mm in total length, with the tail 8 to 15% of the total length. It is a pale tan, pink, or grey snake with 18-60 transverse, brown blotches. The rostral scale is exceptionally thick and turns up and over the snout; the edges are free. A dark transverse mask comes over the crown and through the eyes. The head is slightly wider than the forebody. The eyes are relatively large and have a vertical pupil. Dorsal scales are in 19 rows at mid-body. The rostral scale of this snake is highly distinctive and will separate it from most other snakes with modified rostral scales.

**DISTRIBUTION.** It ranges from Southern Nevada, southeastern California, and southwestern Arizona, southward to the southern tip of Baja and Sinaloa, Mexico. It is associated with creosote bushes and catclaw on sandy, gravel soils.

**DIET.** A dietary specialist, it feeds only on the eggs of lizards and snakes, but will feed on adult lizards, often banded geckos. Sometimes it only gets the gecko's tail.

Skin secretion of this snake suggest that it and the banded gecko have evolved a predator-prey relationship over a long time.

Experimental evidence that the Variegated Banded Gecko, *Coleonyx variegatus*, recognizes the odor *Phyllorhynchus decurtatus* as a predator came from Dial et al. (1989). They investigated the ability of *Coleonyx variegatus* to identify chemicals from the skin of a *Phyllorhynchus* by the geckos' response with a predator-specific tail display. Geckos were presented with five stimuli, including skin chemicals from *Phyllorhynchus decurtatus* and a non-lizard-eating snake, the Mojave Shovel-nose Snake, *Sonora occipitalis*. The skin chemicals from *P. decurtatus* elicited defensive attacks from the geckos in 13% of the trials; tail display occurred in 79% of the trials only in response to skin chemicals from *P. decurtatus.* Thirty-seven percent of the trials involving tail display included rapid flight behavior immediately after the tail display. None of the trials with skin chemicals from the shovel-nosed snake resulted in tail displays from the geckos. However, 30% of these trials resulted in flight. The authors concluded that *C variegatus* can detect snake integument molecules and identify the chemicals' source as a predator. This ability may allow geckos to avoid snakes when foraging

**ACTIVITY.** It is seasonally active from March through November. During the winter, it uses mammal burrows for hibernation, and during the active season it spends its days close to the surface. It is nocturnal. Activity occurs at temperatures in the 28 to 30°C range. At higher temperatures, 36°C and above, it becomes uncomfortable and probably stays below ground.

**REPRODUCTION**. Sexual maturity in males is reached at 223 mm and in females at about 293 mm SVL. Females produce small clutches of 2 to 6 eggs in June and July.

Other snakes have been reported as predators of this species. Its defensive behavior is similar to the Saddled Leaf-Nosed Snake and includes releasing anal musk.

**TAXONOMY and SYSTEMATICS.** Five subspecies have been recognized, but Gardner and Mendelson (2004) found the subspecies did not follow the variation seen across its range and the subspecies have been abandoned.

Figure 84-85. Spotted Leaf-Nosed Snake. Photography JCM

# Hook-Nosed Snakes

*Gyalopion* is a genus of small colubrid snakes with two species native to the southwestern United States and Mexico. One species, *G. canum* is a Chihuahuan Desert form, the other, *G. quadrangulare* is a Sonoran Desert form, both occur in the focus area.

, this will startle a predator and allow the snake to escape.

---

## Thornscrub Hook-Nosed Snake

***Gyalopion quadrangulare*** (Günther, 1893) Figure 86.

**IDENTIFICATION**. Adults reach 354 mm total length with a tail that is 12 to 15% of the total length. The snout is pointed and upturned with a sharp edge. The body is short and stout. The smooth dorsal scales are in 17 rows and have apical pits. There is a single preocular scale, a single loreal, and the anal plate is single. The dorsum has 16 to 41 black, rectangular blotches on the body and another 3 to 10 on the tail. The background color is cream, white, or grayish-white in the mid-dorsal region and red or orange on the sides. The dark blotches are usually narrow on the sides and contain pale spots. The venter is cream to green, yellow, and typically uniform. This is Arizona's most spectacularly colored snake and is difficult to confuse with any other species.

**DISTRIBUTION**. The Thornscrub Hook Nosed Snake is distributed southward from Santa Cruz County in southeastern Arizona to Nayarit, Mexico. It has been found in the focus area near Patagonia, near Peña Blanca Canyon, southward to the city of Nogales, and west of I-19 on Ruby Road.

**HABITAT.** Arizona populations inhabits rolling terrain with mesquite-invaded semi-desert grassland but upslope into adjacent oak savanna at about 1,035 to 1,338 m ASL. In Mexico, the Thornscrub Hook-Nosed Snake inhabits creosote shrublands, thornscrub, thorn-forest, dry tropical forest, and subtropical forest communities.

**DIET.** Prey includes scorpions, spiders, insects and centipedes. Interestingly, only one species of scorpion was found in this snake's digestive system the Ruby Rock Scorpion, *Diplocentrus spitzeri,* despite the fact that at least four other sympatric scorpion species occur with the this snake in Arizona (Babb et al. 2005).

**ACTIVITY.** Seasonal activity is reported from June to November. Daily sctivity is considered crepuscular and nocturnal. The diet is composed of insects, spiders, and scorpions. However, captives reportedly eat other snakes.

**REPRODUCTION**. Females lay clutches of 3 to 6 eggs in July. The hatchlings are patterened like adults.

Figure 86. Thornscrub Hook-Nose Snake. Photography by Yinan Li.

## Chihuahuan Hook-Nosed Snake

***Gyalopion canum*** (Cope 1861) Figure 87.

**IDENTIFICATION.** Adults reach 384 mm total length. The snout is pointed and upturned, and the scales are smooth and in 17 rows. There is a loreal, a single pre-ocular scale, and the anal plate is divided. It has a dorsal pattern with 25 to 52 dark brown blotches or saddles and another eight to 15 on the tail. The blotches usually do not reach the ventrals. The body is gray or gray-brown and may have mottling between the transverse bands. A dark mask extends across the crown and through the eyes, posterior to the snout. Ventral surfaces are cream or white. Males cannot be reliably distinguished from females externally using anatomy because females have small, rudimentary hemipenes. In overall appearance, it may be mistaken for a hog-nosed snake (*Heterodon*). But hognose snakes are larger and has keeled scales. The lack of red, white and black in the pattern will separate it from the Thornscrub Hook-Nosed Snake.

**DISTRIBUTION.** It ranges from west Texas to southeast Arizona and southward into Mexico as far south as Zacatecas and San Luis Posti. The Chihuahuan Hook-Nosed snake occurs in valleys and bajadas of the Chihuahuan Desert, as well as semi-desert grassland, oak woodland, or montane savannas. The high-elevation record is from an area of oak woodland in the rugged terrain of Big Casa Blanca Canyon in the Santa Rita Mountains. There is a Gila County record from an area of transition from Semi-desert Grassland to Inland Chaparral.

**ACTIVITY.** They are commonly encountered after light rain. Activity is primarily nocturnal, but it has also been found during the day.

**DIET.** The diet is composed of spiders, scorpions, and centipedes. Prey is subdued with venom injected via the rear fangs.

**REPRODUCTION.** Males have a seasonal testicular cycle with sperm production occurring between June and September. Females have been reported to produce clutches of one and four eggs in June and July; it is likely that not all females produce each year.

**DEFENSIVE BEHAVIOR.** The Chihuahuan Hook-nosed Snake has an unusual defensive behavior that involves extruding and contracting the lining of their cloaca so that it makes a popping noise.

Figure 87 **Chihuahuan Hook-Nosed Snake. Photography JCM**

# Black-Headed Snakes

The black-headed snakes of the genus *Tantilla* comprises about 64 western hemisphere snakes with flatheads covered with a black cap. They feed on centipedes, scorpions, and other small arthropods. They are often found under rocks and may be buried in the soil beneath them. Their grooved rear fangs deliver venom to the prey during the bite.

## Southwestern Black-Headed Snake

*Tantilla hobartsmithi* Taylor, 1937 Figures 88-89.

**IDENTIFICATION.** A small (8-15 cm in length), gracile snake with a flat head and smooth scales in 15 rows, the blackhead snake has 124 to 172 ventrals and 47 to 74 subcaudals. The mental scale contacts the chin shields, a trait that will separate this snake from other Arizona species. This species is most easily confused with *T. nigriceps*. But *T. nigiceps* is not found in the area

**DISTRIBUTION.** Its distribution comprises disjunct populations from southern Cali-

Figure 88-89. Southwestern Black-headed Snake. Photography JCM

fornia to Texas and southward into western Sonora, eastern Chihuahua, northern Coahuila, and western Nuevo León, Mexico. It ranges from sea level to 1,980 m ASL.

**HABITATS**. Frequently found in riparian areas running through grasslands, chaparral, creosote, mesquite, pinyon-juniper woodlands, and coniferous forests.

**ACTIVITY.** It is seasonally active from March to September. Daily activity is primarily nocturnal but has been found during the day.

**DIET.** Invertebrates with elongated bodies, including soft-bodied caterpillars, beetle grubs, scorpions, millipedes, and centipedes.

**REPRODUCTION**. The smallest sexually mature male was 13.8 mm, and the smallest sexually mature female was 14.1 mm SVL. Females lay clutches of 1 to 3 eggs between May and August. Smith's Black-headed Snake has been found in association with the Variable Sand Snake, *Sonora cincta*.

---

## Chihuahuan Black-Headed Snake

*Tantilla wilcoxi* Stejneger, 1902 Figure 90.

**IDENTIFICATION.** Adults grow to 355 mm in total length, with a tail that is 24 to 26% of the total length. The dorsum is light tan to brown and in contact with the posterior edge of the parietal scales. The wide collar will separate this species from all other Arizona *Tantilla*. The anterior border of the light collar crosses the parietal scales. The collar is bordered by a thin gray-brown band that may be interrupted and form a row of spots. The upper labials are primarily white and extend to the side of the head to create a pale cheek patch. Tiny dark spots are present on the dorsal scales. The venter is cream or white anteriorly, becoming light pink or orange posteriorly.

Figure 90. Chihuahuan Black-Headed Snake. Photography Larry L C. Jones

**DISTRIBUTION and HABITAT.** The species extends from southeastern Arizona southward to Nuevo León, Mexico. Arizona's population uses rocky slopes covered with semi-desert grassland and pine-oak forests between 1,545 and 1,875 m. The habitat has also been described as Warm-Temperate Thorn Steppe. This is the most specialized habitat used by Arizona *Tantilla*, although it has been reported from more arid and lower elevation habitats outside of Arizona (914 to 2,438 m).

**DIET.** Like other *Tantilla,* it feeds on arthropods.

**REPRODUCTION.** Females lay clutches of 1 to 3 eggs.

**PREDATORS.** Predators include night snakes of the genus *Hypsiglena* and ringneck snakes of *Diadophis*.

---

### Yaqui Black-Headed Snake

*Tantilla yaquia* Smith, 1942 Figure 91

**IDENTIFICATION.** Adults grow to 325 mm in total length. Hatchlings are 104 to 140 mm; tails are 21 to 27% of the total length. The dorsum is gray or tan with a dark gray head cap. The posterior edge of the head cap has a white border. The head cap extends two to four scales behind the parietals. The cap extends along the sides of the neck behind and below

Figure 91. The Yaqui Black-headed Snake. Photography Brad Wilson.

the corner of the mouth. Upper labials 5 and 6 are all white, a trait that will separate it from all other Arizona *Tantilla*. The ventral surface is cream anteriorly, gradually becoming pink or orange posteriorly. The scales are smooth and in 15 rows.

**DISTRIBUTION.** The species occurs from near sea level to 1,680 m on the Pacific versant, from southeastern Arizona and extreme southwestern New Mexico, United States, to Nayarit, Mexico. In the United States, it occurs from Cochise and Santa Cruz counties in Arizona and Hidalgo County, New Mexico, southward to Nayarit. In the focus area, it is known from both the Santa Rita and Patagonia Mountains,

**HABITAT.** In Arizona, it occurs in oak woodlands and occasionally in semi-desert grasslands, Chihuahuan desert scrub, and pine-oak woodland. The habitat it uses has also been described as Warm-Temperate Thorn Steppe. This species may use Riparian corridors based on one specimen removed from a Tarahumara Frog's stomach. It is found on the ground under rocks and other debris after rain. Like other black-headed snakes, it probably feeds on invertebrates. Females produce small clutches of eggs, probably one to four per clutch.

## Ground Snakes

The genus *Sonora* contains 14 species of small, rear-fanged snakes that feed on invertebrate prey and range from the central and western United States and northern Mexico. Studies examining the diet of these snakes found spiders, scorpions, centipedes, windscorpions (Solifugae), and insects, including adult and larval beetles, ants, moths, and crickets. Some of these prey species are either dangerous or difficult to capture and consume due to their shapes, exoskeletons, or venom. Many, if not most of these small snakes spend time below the surface and in some cases are "sand swimmers."

## Ground Snake

*Sonora semianulata* Baird and Girard 1853 (?) Figures 92-93.

**IDENTIFICATION.** Adults reach 460 mm, and hatchlings are 70 to 127 mm in total length. The head is not much wider than the neck. The body is cylindrical with smooth scales in 15 rows. This is a highly polymorphic species: some specimens are uniform in color, others are

Figure 92- 93. The Ground Snake. Photography by JCM.

Coral Snake mimics, and others are spotted or striped. Coloration and patterns can be very diverse at one locality. Almost all morphs have a dark spot or bar on the anterior edge of each scale, a trait that will distinguish it from all the shovel-nosed snakes and other small, unrelated snakes. Some specimens have patterns that mimic Black-Headed Snakes (*Tantilla*). The Western Ground Snake has a loreal scale, but black-headed snakes lack a loreal. This species may also be mistaken for Shovel-Nosed Snakes, which have shallower, flatter heads and countersunk lower jaws. The divided subcaudal scales will separate this species from small specimens of the Long-Nosed Snake, which have single subcaudal scales.

**DISTRIBUTION.** The species ranges from southwestern Missouri to eastern Oregon and

California and southward into Mexico as far south as Tamaulipas. The distribution is disjunct with many isolated populations in the northern part of the range. In Arizona, it occurs over most of the western and southern portions of the state but avoids extreme deserts unless it follows riparian corridors. In the focus area, this snake is not common, and it is included here based on observations of knowledgeable individuals.

**HABITATS.** tend to be xeric or semi-arid and at elevations under 2,000 m. Vegetation may be grasslands, deserts, scrublands, or riparian with willows and oak-juniper. It is often found in areas with moist soil.

**ACTIVITY.** is primarily nocturnal, but it may also occur during the day. Activity may increase after rain, and seasonal activity may extend from late February to November.

The diet includes a variety of invertebrates but frequently consists of spiders, solpugids, and orthopterans. It has rear fangs and venom, but is harmless to humans.

**REPRODUCTION.** Males mature at about 183 mm, and females at about 265 mm. Females lay clutches of three to six eggs in June or July, probably timed with the onset of the monsoons. This small snake undoubtedly has many predators, including some invertebrates, such as the giant centipede, *Scolopendra heros*.

**TAXONOMY.** Baird and Girard (1853:117) originally gave the type locality as "Sonora, Mex." However, it has been restricted to the Santa Rita Mountains, Arizona. The type locality in the description was "Sonora, Mexico." Stickel (1943) restricted the type locality to the vicinity of the Santa Rita Mountains of southern AZ, based on historical records from the expedition. However, this locality was in the Mexican state of Sonora prior to the Gadsden Purchase of 1854.

---

## Tucson Shovel-Nosed Snake

*Sonora annulata klauberi* (Stickel 1941) (?) Figures 94-95.

**IDENTIFICATION.** Adults grow to 430 mm total length, with a tail that is 17 to 20% of the total length. The head is the same width as the body. It has a snout flattened and spade-like, a countersunk lower jaw, and valvular nostrils. It has a wide black hood or band on the crown with a concave anterior edge. The dorsum has smooth, shiny scales. It has a cream or yellow body with fewer than 21 narrow, transverse black bands that usually don't completely encircle the body. They are separated by red cross bands that are often incomplete and indistinct. Another four or five black bands are on the tail. Fifteen dorsal scale rows at mid-body will separate it from the Variable Sand Snake.

**DISTRIBUTION** extends from extreme southeastern California southward into Baja California and northern Sonora, Mexico, and east into southwestern Arizona. It has two subspecies. The Colorado Desert Shovel-Nosed Snake, *C. a. annulata*, is found in extreme southeastern California, east of the desert slope of the peninsular range to the Colorado River. It ranges southward into Baja California and northern Sonora, Mexico, and eastward into southwestern Arizona. The second subspecies is the Tucson Shovel-Nosed Snake, *C. a. klauberi,* that Wood et al. (2014) found in central and western Arizona, including La Paz, Maricopa, Pima, Pinal, Yavapai, and Yuma counties.

**ACTIVITY** Nocturnal, it is a sand-swimming species that spends most of its time below the surface. It may occasionally be found by day in shaded areas, but it is most often found crossing desert roads at night. Deserts with little vegetation, loose soil, and rocky regions are preferred habitats.

Active foraging is commonly used to locate food. This species' diet is composed of arthropods, and it may prefer scorpions. Centipedes, spiders, lepidopteran pupae, beetles and their larvae, and lizard eggs are also consumed. Prey may be captured and swallowed while the snake is underground.

**REPRODUCTION**. Males mature at about 201 mm SVL, and most are mature at 250 to 260 mm TL. Copulation has been observed in the field in late May. Clutches of usually 2 to 6 eggs are laid from March to July. Incubation has been reported to be 94 to 99 days. Hatching probably occurs from June to September. Hatchlings are 112 to 124 mm TL. Male combat in captive males involved snakes intertwined and writhing under the sand for less than a minute. Biting, entwining, and tail vibration were also observed. Scales are 182 to 216, subcaudal scales are 120 to 167. The Striped Whipsnake usually has 15 dorsal-scale rows and is darker in color.

This snake ranges from central Arizona and southwest New Mexico southward to Jalisco

Figure 94-95. Tucson Shovel-Nosed Snake. Photography Timothy A. Cota

and Colima, Mexico. It is also present on some of the islands in the Gulf of California. It is frequently associated with rocky canyons and forested, steep mountain terrain but may also occur in rocky deserts and relatively flat terrain along riparian corridors. It may reach elevations of 2,200 m, but also uses creosote flats at 1,200 m.

## Variable Sand Snake

*Sonora cincta* (Cope, 1861) Figures 96-97.

**IDENTIFICATION**. The dorsal pattern is composed of transverse black bands between yellow, orange, or red rings, or it can be a uniform dorsum with each dorsal scale having a brown spot (in the Baja). The ventral surface is cream to yellow with black bands encircling the body. A fossorial snake, it has small eyes, lacks a loreal scale, has a short tail and roads at night. Deserts with little vegetation, loose soils, and the rocky regions seem to be preferred habitats.

**DISTRIBUTION**. The Variable Banded Sand Snake is distributed in the Sonoran desert and dry forests of southern Arizona, western Sonora south to Sinaloa and on the northern Baja California Peninsula at least as far south as Bahia Concepcion. They have been found from sea level to 1000 m ASL.

**FORAGING** and **DIET**, Active foraging is commonly used to locate food. The diet is composed mostly of arthropods, and it may prefer scorpions. Also consumed are centipedes, spiders, lepidopteran pupae, beetles and their larvae, and lizard eggs. Prey may be captured and swallowed while the snake is underground.

**TAXONOMY**. Note that this snake was long considered to be in the genus *Chilomeniscus* and was known as *Chilomeniscus stramineus* in Arizona. Variation in *Sonora cincta* scales was summarized in Holm (2008). Individuals can be either uniform or with red, or orange and black or brown crossbands. The bands can be complete and cross the ventral scales, or saddles that reach the lateral dorsal surface, or dorsal spots. There is also variation in pre

Figure 96-97. A Variable Sand Snake responding to the ordor of a kingsnake on the photographer's hand. The unusual posture now known as body bridging was first reported as an unusual defensive behavior in rattlesnakes (genus Crotalus) to the presence of kingsnakes (Lampropeltis getulus), notorious predatory snakes on rattlesnakes. Cowles (1938) called it the "kingsnake defense posture" where the kingsnake elevated the middle portion of the body to form a bend (body bridge), which directs body blows toward an advancing kingsnake, a decided contrast to the vertical coiling position assumed by rattlesnakes to other types of predators. The behavior was shown to be triggered by chemical cues from the cloacal glands of the kingsnake. It has now been found in colubrid snakes that often fall prey to kingsnakes.

Figure 98-99-. Variable Banded Sand Snake. Photography JCM

**REPRODUCTION**. Males mature at about 194 mm SVL, and most are mature at 250 to 260 mm TL. Copulation has been observed in the field in late May. Clutches of usually 2 to 6 eggs are laid from March to July. Incubation has been reported to be 94 to 99 days Hatching probably occurs from June into September. Hatchlings are 112 to 124 mm TL.

Male combat in captives involved snakes intertwined and writhing under the sand for less than a minute. Biting and tail vibration were also observed.

## Whipsnakes and Coachwhips

Snakes in the genus *Masticophis* are called whipsnakes or coachwhips, which are endemic to the Americas. They are characterized by having a long, thin body and are not dangerous to humans.

Coachwhip snakes' diet consists primarily of lizards, other snakes, insects, birds and their eggs, and infrequently various amphibians. Researchers have documented instances where coachwhips have consumed rodents and carrion, but this is rarely the case. Coachwhip snakes are a foraging species and actively search for their prey.

Right after hibernation, coachwhip snakes are active primarily in midday. As their active season progresses, they increase their time spent moving, preferring mid-mornings and late afternoons for foraging. Coachwhip snakes move an estimated 186 m/day, moving from vegetated areas to burrows, foraging for their prey.

Coachwhips rely on visual and olfactory senses to capture their prey. They can perceive chemicals emitted from prey, precisely distinguishing prey species from those that are not prey. Coachwhips flick their tongues and present attack behaviors when they process chemicals from their preferred prey. Their most common prey, lizards and snakes, elicit the strongest response. Once the prey has been located, coachwhips rely on their speed to capture the animal. As the fastest snake on the North American continent, with a maximum speed of 5.8 kph, capturing prey is made less complicated.

---

### Sonoran Whipsnake

*Masticophis bilineatus* Jan, 1863. Figures 100-101.

**IDENTIFICATION.** This snake has also been called *Coluber bilineatus*. Adults grow to about 1,700 mm with a tail about 40% of the total length. This is a long, slender snake with a head wider than the forebody. The dorsum is a gray-green, brown-green, or brown-gray with two or three laterals, white to cream stripes on each side of the body. The stripes become indistinct posteriorly. An interrupted short black line separates the stripes. Dorsal scales are smooth, in 15 rows at mid-body, with an apical pit, and a pair of pale spots on the anterior of the scale. Ventral scales are 182 to 216, subcaudal scales are 120 to 167.

**DISTRIBUTION.**This snake ranges from central Arizona and southwest New Mexico southward to Jalisco and Colima, Mexico. It is also present on some of the islands in the Gulf of California. It is frequently associated with rocky canyons and forested, steep mountain terrain but may also be found in rocky deserts and relatively flat terrain along riparian corridors. It may reach elevations of 2,200 m, but also uses creosote flats at 1,200 m. It is associated with the communities of the Madrean Basin Grasslands, Madrean Lower Montane Woodlands, Lower Mogollon Transition, Arizona Upland/Eastern Sonoran Mountains, and Low Mountains and Bajadas (Fleischner et al. 2024).

**FORAGING AND DIET.** The Sonoran Whipsnake is a diurnal hunter and feeds on frogs, lizards, birds, and small mammals, including bats. While it tends to be ground-dwelling, it does not hesitate to climb trees. Sexual maturity is reached at about 760 mm. Arizona females collected in June and July laid 6 –13 eggs per clutch.

**PREDATORS.** This species is eaten by raptors, carnivorous mammals, and the rock squirrel, *Spermophilus variegatus*.

Figures 100-101 Sonoran Whipsnake. Top photo 91 is an adult preparing to shed its skin. Photography JCM. Bottom 92 is an adult foraging for food on the ground  Photography by National Park Service.

## Western Coachwhip

*Masticophis piceus* (Cope 1892) Figures 102-104.

**IDENTIFICATION.** Adults grow to 2,590 mm in total length, with the tail 21 to 28% of the total length. Hatchlings are 205 to 240 mm in total length. A slender, fast-moving snake with smooth scales, it has dorsal scales in 17 rows, reduced to 13 rows in front of the vent. Ventral scales in males average 193 (189-197) and females average 195 (185-205). Each scale has two apical pits. Head and eyes are large. The tail has a braided appearance. Contrary to O'Connell and Smith (2017), my observations in Pima County, west of the Cochise Filter Barrier, suggest that black individuals are not in isolated areas but live right next to red specimens. Neck bands are common in some areas, and some have dark red-brown

Figures 102-104. *Masticophis piceus*. top 102 a juvenile periscoping; middle 103; is a black and partial red adult; bottom 104 is a red-banded adult. Photography JCM.

ground color on upper surfaces, broken by complete transverse, narrow, light-colored pink cross-bands. The populations in Cochise County, east of the Cochise Filter Barrier, are also understudied.

**HABITAT**. The Western Coachwhip is a grassland, savanna, desert, and scrubland species that ranges from the San Joaquin Valley in California to western Nevada, southern Utah, and western Arizona in Mexico, northern Baja California, Sonora, and Sinaloa. Its elevation ranges from sea level to 2,300 m.

**DIET.** The diet includes most vertebrates, and it will follow prey into trees or burrows.

**REPRODUCTION.** Sexual maturity is reached at 700–900 mm. Clutches of 4 –20 eggs are laid in June and July, and hatch in August and September after an incubation period of 44 –88 days. Hatchling Coachwhips have a series of dark transverse bars on their dorsum.

**THERMAL RELATIONSHIPS.** Coachwhips have some of the highest body temperatures recorded in snakes. When they emerge in the morning from their burrow, they can have a body temperature of 21C. After basking in the sun, they have been recorded to have temperatures as high as 47.5C — more than 10C higher than human body temperature— but they tend to average 33–34C.

**DEFENSE**. Coachwhips are one of the few snakes that may advance toward a human when disturbed. If grabbed, they do not hesitate to bite and chew. O'Connell et al (2017) recognized this species based on molecular data.

**TAXONOMY**. It is composed of the former subspecies *M. f. pices, M. f. cingulum*, and *M. f. ruddocki*,

Figure 105. A black morph of *Masticophis pices.* Photography JCM

# Green Rat Snake

*Senticolis* is a genus of snake in the family Colubridae. The genus is monotypic, containing the single species *Senticolis triaspis,* also known as the Green Rat Snake. The species is endemic to Central America, Mexico, southern Arizona, and southern New Mexico.

It is of interest because it has been shown to be the sister to the other members of the Lampropeltini (the kingsnakes and their relatives).

---

## Northern Green Rat Snake

*Senticolis triaspis intermedia (*Boettger, 1883) (LC) Figures 106-108.

**IDENTIFICATION.** Adults reach 1,600 mm, with the tail about 25% of the total length. Females are larger than males. Hatchlings are 280 to 400 mm. The head is long and broader than the forebody. The dorsum in adults is a uniform green, yellow-green, or olive, with a cream to yellow ventral surface. The dorsum in young individuals is gray to gray-green with gray-brown transverse blotches that disappear with age. The dorsal scales are weakly keeled dorsally and smooth laterally. The scales have apical pits and are usually in 33 (range: 30 to 39) rows at midbody. The ventrals are keeled (the outer edge forms a 90-degree angle) and number 241 to 282. There are 85 to 126 subcaudals. The head is relatively long and narrow and is clearly distinct from the thin neck. The pupils are round. The combination of the high number of dorsal scale rows and the uniform green coloration in adults distinguishes this species from all other Arizona snakes.

**DISTRIBUTION.** The subspecies is the Northern Green Rat Snake, *S. t. intermedia* (Boettger, 1883). Its distribution ranges from southeast Arizona and southwest New Mexico southward through Mexico to Costa Rica. In southeastern Arizona, it occurs in the Baboquivari, Pajarito, Santa Rita, Empire, Whetstone, and Chiricahua mountains. It occurs from near sea level to 2,500 m ASL.

**HABITATS.** Madrean Woodland, with Desert Scrub influences and chaparral in rocky montane canyons, are often used and it is often found on east-facing slopes near streams. Its secretive nature gives the impression that it is uncommon, but it likely spends much of its time in crevices and burrows and is not frequently encountered. One radiotelemetry study found that Green Rat Snakes moved most often in the afternoon. Males may show a preference for riparian habitats, while females may prefer desert scrub.

During radio tracking they were secretive, hiding in loose talus, vegetation, wood debris, and rodent burrows (Radke and Malcom, 2008).

**ACTIVITY.** Seasonal activity in Arizona is from May to October. Daily activity tends to be diurnal and crepuscular, but snakes will shift to nocturnal activity during the heat of the summer. Snakes that were monitored for body temperature maintained a temperature of 25°C, relatively low for a Sonoran Desert snake. Radiotracked snakes were active at body temperatures of 23.4 to 29.7 C (Radke and Malcolm 2005). The species will climb,but tends to be terrestrial.The species is secretive and may be more abundant than is commonly thought.

**DIET.** Constriction is used to kill prey. Wild prey includes woodrats, mice, bats, birds, and lizards. Captives presented with a variety of prey only took mammals

**REPRODUCTION.** Females are mature at 625 mm and about two years of age. Clutches of 3 to 9 eggs have been reported; they require an average of 53 days of incubation. Hatching occurs in late August. The reproductive cycle is thought to be aseasonal and continuous.

Figures 106-108. Top is a juvenile Green Rat Snake, middle 107 is a normal patterened adult, bottom image 108 shows the head of a large adult. The Green Rat Snake, is known for its green or olive-green dorsal coloration and light cream or yellow ventral side. Juveniles have gray to gray-green coloration with brown blotches, which fade as they mature. Photography JCM

# The Neotropical Treesnake Clade

One species of Arizona snake had its ancestor in a clade of South American tree snakes. The Thornscrub Vine Snake seems out of place in North America; however, it reaches the most northern edge of its range here. The genus is related to the brightly colored, arboreal parrot snakes in *Leptophis*. The Thornscrub Vine Snake, *Oxybelis microphthalmus,* ranges from Arizona to Oaxaca and in found only on Mexico's west coast. in north-central Sonora.

---

## Thornscrub Vine Snake

*Oxybelis microphthalmus* Barbour and Amaral 1926. Figure 109.

**IDENTIFICATION.** This is perhaps the most easily identifiable snake in Arizona. with its small diameter body and elongated head. Adult males vary in total length from 1197-1337 mm, SVL varies from 713–834 mm and tail lengths vary from 484–538 mm; tail/SVL ratios vary from 0.63–0.74. Female total lengths vary from 667–1407 mm; tails vary from 250–544 mm. Tail/SVL ratios in females vary from 0.60–0.72. Ventrals in males vary from 184–202 ($n = 14$, $x = 192.0$, $SD = 5.92$); ventrals in females vary from 184–204. Subcaudals in males vary from 163–175; in females; subcaudals vary from 170–183.

The dorsum is gray to brown, the crown of the head is brown to bronze with a yellow to white lateral-ventral surface that includes the upper and lower labials. A thin line of black pigment separates the upper brown-bronze coloration from the lower white to yellow pigment. Dorsal scales are arranged in 17 rows , Arizona populations have 192-197 ventrals.

**HABITAT.** In southern Arizona and northern Sonora, the habitat includes desert-grassland, and pine-oak woodland communities at 1160 to 1650 m elevation. The restricted area occupied by *0. microphthalmus* in southern Arizona is characterized by relatively mild

Figure 109. Thornscrub Vine Snake. Photography JCM

winter temperature and relatively high annual precipitation that falls mainly as summer monsoon rain. The northerly limit of *Oxybelis* in southern Arizona is apparently limited by winter freezing above 1650 m and by summer aridity below 1160 m.

**DIET.** Lizards make up most, if not all of the prey of the Thornscrub Vine Snake. One specimen was killed and collected with a Lesser Earless Lizard in its mouth.

**REPRODUCTION.** Goldberg (1998) found males produce sperm from February through July. Two clutches of four eggs have been reported from Arizona.

# Patch-Nosed Snakes

Patch-nosed snakes of the genus *Salvadora* are slender, diurnal snakes with an enlarged and thickened rostral scale covering their snout. These snakes forage for reptile eggs and dig them out of loose soil. However, they will also eat lizards, snakes, and small mammals. Patch-nosed snakes are related to the racers and whipsnakes. However, the patch-nosed snake genus *Salvadora* was not monophyletic in a recent snake phylogenic study that used molecular data (Figueroa et al. 2016) and was monophyletic in a study that uses morphology (Hernández-Jiménez et al. 2021). The tropical Amazon Banded Snake, *Rhinobothryum lentiginosum*, was the sister to the Mountain Patch-nosed Snake and the Mexican Patch-Nosed, *Salvador mexicana*, while the Western Patch-Nosed was the sister to the Lyresnake, *Trimorphodon biscutatus* in the molecular study and the sister was is in a clade that forms the sister to the gophersnakes. This raises the question of whether these two similar species are related to each other or whether they belong to separate clades. Or do members of the the genus have rear fangs and venom glands as well as a similar appearance because of convergent evolution? I follow Hernández-Jiménez et al. (2021) and expect name changes in the future once the genus is well represented with molecular data.

In Appendix 5 are characters that are used to identify patch-nose snakes Figure 100 has been adapted from Hernandez-Jimenez et al. (2021).

---

## Mountain Patch-Nosed Snake

*Salvadora grahamiae grahamiae* Baird & Girard, 1853 Figure 110.

**IDENTIFICATION.** Adults reach 1,194 mm total length; hatchlings are 216 to 290 mm in total length. Males are smaller. The dorsum is grey to green-grey or light brown. Two dorsolateral stripes extend from the eye along the length of the body. In Arizona specimens, the lateral stripes on the third scale row are often faint or absent. The area between the stripes is brighter in color than the rest of the snake. The venter is a uniform cream, or white which separate it from the Big Bend Patch-Nosed Snake and the Western Patch-nosed Snake.

**DISTRIBUTION.** The Eastern Patch-nose Snake ranges from Northern New Mexico and Central Arizona southward into Sonora, Chihuahua, and Coahuila, Mexico.

**HABITAT.** The Mountain Patch-Nosed Snake prefers mountainous terrain above 1,200 m. Thus, it occurs in mountains, mesas, arroyos, and canyons with thornscrub, oak-juniper, and oak-elm woodlands. It is associated with the Sky Island archipelago.

**FORAGING and DIET.** Basking and foraging for prey usually occurs in the morning. They can be seen moving about, inspecting crevices, burrows, and piles of vegetation. Like their relatives, these are active hunters, a behavior distinguishing them from other snakes that hunt primarily from ambush. They feed on lizards, small mammals, and reptile eggs. The thick, patch-like rostral scale protects its soft tissue while it excavates reptile nests for eggs. This species has been found to have venom that causes hemorrhages and muscle necrosis and is probably used to prepare prey for digestion. It is thought to be harmless to humans.

**REPRODUCTION.** Males may reach maturity at 480 mm, and females at 560 mm. Clutch sizes up to nine have been reported in eastern populations. Eggs are laid from April to July, and hatchlings are present from July to October.

Figure 110. Mountain Patch-Noseed Snake. Photography Laurie Vitt

## Western Patch-Nosed Snake

***Salvadora hexalepis*** (Cope, 1866)

**IDENTIFICATION**. Adults reach 1,190 mm in total length; the tail is about 22 to 23% of the total length; hatchlings are about 250 mm in total length. Males are larger. The eyes are large and the top of the head is grey. The vertebral stripe is three scales wide. The dorsum is yellow to tan with two dark brown stripes, each divided into a wider upper stripe and a narrow lower stripe on scale-rows three or four. The region between the stripes is only three scale-rows wide and may have faint cross bands. Dorsal scales are smooth with apical pits, and in 17 to 19 rows. Ventral scales number 187 to 215, and subcaudals number 73 to 99. There are nine upper labials. The loreal is divided. The rear pair of chin shields are separated by two or three small scales. The combination of these traits will separate it from the Big Bend Patch-Nosed Snake and the Mountain Patch-Nosed Snake.

**DISTRIBUTION**. The Western Patch-Nosed is relatively widespread, ranging from southern California, Nevada, extreme southwestern Utah, southward across Arizona, southern New Mexico, and southwestern Texas, and southward into Mexico to the tip of Baja California, southern Sinaloa, and Chihuahua. Its elevational range extends from below sea level to 2,135 m ASL.

**HABITAT.** Western Patch-Nosed snakes occupy habitats with flat topography and sandy or rocky soil, lower mountain slopes, creosote bush and ocotillo-yucca-agave flats, mes-

Figure 111. Western Patch-Nosed Snake. Photography JCM

quite-dominated washes, grassland, broken upland terrain, succulent desert, partially dry streambeds of foothills and mesas, sagebrush semi-desert, chaparral, and pinyon-juniper woodland. It is ground-dwelling, but has been found climbing in low vegetation.

**TAXONOMY**. Three subspecies occur in Arizona. The Big Bend Patch-Nosed Snake, *Salvadora hexalepis deserticola*, inhabits upper elevations of Sonoran desert scrub, semi-desert and plains grassland, Chihuahuan Desertscrub, and oak woodlands and savanna. The Desert Patch-Nosed Snake *S. h. hexalepis,* occupies Sonoran desert scrub. The Mojave Patch-Nosed Snake, *S. h. mojavensis,* uses scrublands with more rugged topography. Despite being widespread and locally common, patch-nosed snakes remain poorly studied.

**DIET and FEEDING**. Sullivan and Weinstein (2017) observed an individual from Maricopa County seize and coil around a Western Whiptail with the lizard in its mouth. The snake held the lizard for three minutes before it closed its eyes and went limp. Western Patch-Nosed Snakes have enlarged rear fangs and venom glands. This was the first report of a patch-nosed snake using venom to subdue prey.

Sexual maturity is reached in males at 468 mm and in females at 584 mm. Females lay clutches of 3 to 10 eggs in June to August, hatching from late August to late October.

## Big Bend Patch-Nosed Snake

*Salvadora deserticola* (Schmidt, 1940)

Adults are 1,143 mm; most are less than 900 mm; hatchlings are 210 to 280 mm. This snake, previously considered a subspecies of *Salvadora hexalepis,* has been elevated to full species status. Bogert (1945) describes characters justifying the distinction and states that there are no intermediates. The following characters from S. hexalepis can diagnose the

Figure 112 Big Bend Patch-Nosed Snake. Photography Laurie Vitt.

species: two upper labials enter the orbit (as opposed to none or only one in the orbit of *hexalepis*), a pale grey coloration (as opposed to yellow-tan) a single loreal, and the narrow lateral line on the fourth scale row instead of on the third and fourth at mid-body.

**DISTRIBUTION**. West Texas, southern New Mexico, and southeastern Arizona, extending southward into Sonora, Chihuahua, and Sinaloa, Mexico. The Big Bend Patch-Nosed Snake is a Chihuahuan Desert species found at elevations between 1000 and 1,600 m ASL. It occurs in terrain with loose soils, rock piles, and a mix of creosote, mesquite, and ocotillo. This is an active foraging snake, hunting lizards, snakes, and small mammals in the morning and late afternoon. It will also likely eat eggs and small birds.

Males are sexually mature by the time they reach 470 mm; females are mature by the time they reach 590 mm. Females produce clutches of 3 to 10 eggs, but average about 6. Egg laying typically occurs in early June, with hatchlings emerging in August and September.Big Bend Patchnose Snake *Salvadora deserticola* (Schmidt 1940).

**IDENTIFICATION.** Adults can attain 1,143 mm; but most are less than 900 mm; hatchlings are 210 to 280 mm. This snake, previously considered a subspecies of *Salvadora hexalepis,* has been elevated to a valid species (Hernández-Jiménez et al. 2021). The loreal is usually single; the vertebral line is five scale rows wide on the first third of the body, and the lateral line is well defined. Subcaudals number 68 to 86.

**DISTRIBUTION** West Texas to southern New Mexico and southeastern Arizona southward into Sonora, Chihuahua, and Sinaloa, Mexico. The Big Bend Patch-Nosed Snake is a Chihuahuan Desert species found at elevations between 1000 and 1,600 m ASL. It occurs in terrain with loose soils, rock piles, creosote, mesquite, and ocotillo. This is an active foraging snake, hunting lizards, snakes, and small mammals in the morning and late afternoon. It will also likely eat eggs and small birds.

**REPRODUCTION**. Males are sexually mature by the time they reach 470 mm, females are mature by the time they reach 590 mm. Females produce clutches of 3 to 10 eggs, but average about six. Egg laying may occur in early June, with hatchlings appearing in August and September.

# The Lyre Snakes

*Trimorphodon* is a genus of seven mildly venomous, rear-fanged, colubrid snakes. They are commonly known as lyre snakes, named after the distinctive V-shaped pattern on the crown behind the eyes. Lyre snakes are nocturnal and can be active in very dry conditions as well as during monsoon rains. They climb well and have been found searching rock crevices for prey. They may use the rock crevices as daytime refugia. Prey is mostly lizards, but it will take birds and mammals. Prior to 2008 the species was known as a subspecies of *Trimorphodon biscutatus*: *T. b. lambda*.

Figure 113. The Sonoran Lyresnake. Photography JCM.

## Sonoran Lyresnake

***Trimorphodon lambda*** Cope, 1886 Figures 113-116.

**IDENTIFICATION.** Adults reach 1,210 mm total length, with the tail about 14% of the total length. Hatchlings are 205 to 264 mm total length. The head is broad and the forebody very slender. There is a white crossband on the snout, a white crossband between the eyes, and a chevron or lyre-shaped mark behind the eyes. This lyre-shaped marking is diagnostic for the species and the source of the common name. The dorsal coloration often matches the rocks and soils of the snake's habitat. The body is grey to light brown with 20 to 28 large, dark brown dorsal blotches with light edges and a pale crossbar in the center. It has lateral irregular blotches between the more dorsal bands. The eyes are relatively large with vertical pupils. Ventral surfaces are near white or yellow with dark spots. Males tend to be shorter and have fewer blotches than females. Smooth dorsal scales, each with a pair of apical pits, are arranged in 15 to 18 rows at mid-body.

**DISTRIBUTION and HABITAT.** The Sonoran Lyresnake's range extends from Nevada southward through Arizona and into Sonora, Mexico. It inhabits rocky slopes and canyons, where it hides in deep fissures in the rocks, but it has been found away from rock outcrops in desert scrublands and forests from sea level to 2,200 m.

Stocking (2023) reports this species from a rocky canyon with desert scrub, a habitat where I have usually found this species. However, in 2024, I removed one of these snakes from the metal track of a garage door in an urban area. What the snake was doing there is unknown.

**DIET.** Lyresnakes feed mostly on lizards, rodents, and bats, and captives have eaten birds. The snake has venom glands and rear fangs. This is one of the few species of snakes that may use both venom and constriction to kill prey. Rodents are held in coils and constricted, while lizards are held in the mouth and envenomated. However, it is possible that the lyresnake simply holds a rodent in its coils while it envenomates it, and does not actually kill it by constriction.

Jones (1971) fed three captive specimens for 15 months. The snakes accepted prey from late March to late October. They were offered mice and lizards, the mice were secured by a strike followed by constriction. The lizards were struck and retained in the snakes' mouth until they were swallowed.

**REPRODUCTION.** Sexual maturity is reached in males at about 448 mm SVL and in females at 711 mm SVL. Females lay clutches of 6 to 20 eggs that require 77 to 79 days of incubation.

**ACTIVITY.** The Sonoran Lyresnake is nocturnal and may be encountered crossing roads. When disturbed, it will flatten its head, raise it body, and strike.

**VENOM.** While the venom is effective on lizards, it does not appear to be dangerous to humans. Emerson (2024) recounts the events of a 2019 bite from this snake. The bite was initiated when a wild individual was picked up. The fangs entered the right forearm and remained for 63 seconds. The snake was released. But within a few minutes, the bite site had alternating numbness and radiating pain. If pressure was applied, a burning sensation was felt at the bite site. Within an hour, the author experienced local discoloration, nausea, sharp abdominal pains, vomiting, and swelling. This bite was unusual because the envenomation led to systemic symptoms. Systemic symptoms are not usually associated with rear-fang snakebites. It's best not to handle these snakes for both the human's and the snake's safety.

Figures 114-116. *Trimorphodon lambda*. Top two images are adults. Bottom image is a hatchling. Photography JCM.

# Kingsnakes & Their Relatives

When I moved to Southeast Pima County in 2015, I was surprised to find black kingsnakes "*nigrita*" living in my neighborhood. But I also found banded kingsnakes and speckled kingsnakes. The black kingsnakes were the most common in 2015, by 2016, the black kingsnakes disappeared, but the speckled and banded pattern individuals remained. Herpetologists have considered all of these color morphs of *Lampropeltis californiae* (Drost in Holycross and Mitchell 2022). However, Pyron & Burbrink (2009) published a map that suggests the Desert Kingsnake (*Lampropeltis splendida*) was primarily restricted to the east of the Cochise Filter Barrier but they suggested some populations may be on the Mollogon Rim and in southeast Arizona. The identity of these populations remains to be determined. The disappearance of black kingsnakes between 2015 and 2016 is difficult to explain if they were just morphs or old individuals of *L. californiae* that had lost their yellow pigment. Nor does it explain why Krysto et al. (2017) found the black kingsnakes "*nigrita*" genetically distinct from their *L. californiae* sister.

Using molecular data, Pyron and Burbrink (2009) found the Lampropeltini diversified about 24 MYA. Generic diversification events occurred in the early to mid-Miocene; speciation events happened before the Pleistocene.

---

## Eastern Glossy Snake

*Arizona elegans elegans* Baird, 1859 (LC) Figures 117-119.

**IDENTIFICATION.** Adults reach 1,780 mm, with the tail 12 to 17% of the total length. Males are smaller than females. The ventral surface is uniform white or cream. It has two prefrontals, a trait that will separate it from its sister, the Gophersnake. The dorsal scales are smooth and in 25 to 35 rows at mid-body. The anal plate is single. The combination of smooth dorsal scales, the high dorsal scale row count, the uniform venter, and the single anal plate will separate this species from other Arizona colubrid snakes.

**DISTRIBUTION.** Widely distributed from Nebraska and Texas to California and southward into mainland Mexico and Baja, it uses deserts, grasslands, and open woodlands.

**DIET.** Prey includes mammals and lizards. It has a relatively long head and large gape, to accommodate large prey. Thus, it can eat horned lizards, kangaroo rats, and birds.

**FORAGING.** The time of day they hunt and the type of prey they hunt varies geographically. The Eastern Glossy Snake is active during the day in the eastern part of its range and is more nocturnal in the western portion of its range. It forages for prey in burrows after dark, and ambushes prey on the surface during the day. The species is cold-tolerant and, has been found active at 14C, but Cowles and Bogert (1944) suggegst 19-20C are more likely the lower limit for activity. Most snakes are active at temperatures of 27C. And, the critical thermal maximum for *Arizona elegans* is likely 43-44C (Cowles and Bogert, 1944).

**REPRODUCTION.** Mating occurs in the spring, spermatogenesis occurs in the summer preceeding mating. The smallest reproductive male was 367 mm SVL; the smallest reproductive female was 562 mm (Ernst and Ernest 2003).There is some evidence that females do not reproduce every year. The first egg laying occurs in the second or third summer. Clutches range from 2-12 eggs and they average 7.5 to 8.5 eggs. Hatchlings appear in late August and mid-September. They measure 210-260 mm in SVL. (Persons and Holycross, in Mitchell 2020).

Figures 117-119. Eastern Glossy Snake. Photography JCM

Figure 120-122. Gophersnake. Photography JCM

## Gophersnake

*Pituophis catenifer* (Blainville, 1835) Figures 120-122.

**IDENTIFICATION.** Adults reach 2,130 mm total length, with a tail that is 10 to 18% of the total length. Hatchlings are 200 to 550 mm in total length. This large snake has a head distinct from the forebody. The rostral scale protrudes slightly from the snout. The dorsal background color is yellow or tan with large quadrangular dark brown blotches or saddles and smaller, more lateral gray-brown spots. The ventral surface is a uniform cream or yellow with paired lateral spots. The combination of keeled dorsal scales, a high dorsal scale-row count (27 to 37), and four prefrontal scales will separate this species from all other Arizona colubrid snakes.

**DISTRIBUTION.** This widely distributed species covers most of North America's western half and ranges southward as far as Veracruz, Mexico. In Arizona, it occurs throughout the state from sea level to about 9,000 feet. However, it appears to be absent from the highest mountains. This distribution occurs in all the state's biomes, making it a habitat generalist. If it happens in the Alpine Tundra Biome, it will likely be there only as a visitor, not a permanent resident.

**ACTIVITY.** Gophersnakes have been found active at all times of the day and night. However, its annual activity pattern is to hibernate in the winter and use the warmer days of spring, summer, and fall to hunt for food, search for mates, and lay eggs. While Gophersnakes are usually found on the ground, they are excellent climbers.

Arizona recognizes two subspecies of gopher snakes. The Sonoran Gophersnake, *Pituophis catenifer affinis,* inhabits most of the state except the northern quarter. The Great Basin Gophersnake, *Pituophis catenifer deserticola,* occurs in the northern quarter of the state.

**DIET.** Gophersnakes are generalized vertebrate predators, but much of their diet is mammals. Prey is killed by constriction. In a study of 2,600 specimens, 1,066 prey were found in gophersnake digestive tracts. Mammals composed 74.8% of the diet, birds made up 8.1%, bird eggs made up 11.9%, and lizards 3.3%. They are primarily diurnal, but can also be active at night.

**TAXONOMY and SYSTEMATICS.** Molecular-clock estimate suggests *Coluber constrictor* and *Masticophis flagellum* last shared an ancestor about 12 MYA. Note that O'Connell and Smith (2017) have reorganized the snakes related to *Masticophis flagellum.*

# Long-Nosed Snakes

The Long-Nosed snakes, of the genus *Rhinocheilus,* inhabit much of Mexico and the south-central and southwestern United States. The three species are most frequently seen at night. These snakes are imperfect coral snake mimics. They are the sister to the genus *Arizona.*

## Western Long-Nosed Snake

*Rhinocheilus lecontei lecontei* Baird & Girard, 1853 Figures 123-127

**IDENTIFICATION.** Adults grow to 1,520 mm, with the tail 15 to 17% of the total length. Hatchlings are 180 to 280 mm. The head is about the same width as the forebody. There is a long, pointed snout and a countersunk lower jaw. This is a slender snake with a highly variable pattern.Usually, its a saddle-like pattern of red, black, and white. Often the white is heavily flecked with black specks in most specimens, and the saddle blotches alternate red and black. The saddle blotches do not intrude on the venter. The belly is a uniform cream

Figures 123-125. Variation in color patterns in the Long-nosed Snake.

Photography JCM

or yellow. Some individuals lack red completely and are black and white. Dorsal scales are smooth and in 23 rows at midbody; the anal plate is single, and the subcaudals are single. It is most likely to be confused with tri-colored kingsnakes that have paired subcaudal scales.

**DISTRIBUTION.** The species is widespread from Nebraska and Texas west to California and southward into Baja and mainland Mexico as far south as Nayarit. Habitats used include lowland deserts, dry prairies, and scrub woodlands (mesquite, acacia, prickly pear cactus) with sandy soils.

**ACTIVITY.** Seasonal activity occurs from March to October, but it has been found in all months of the year. It is crepuscular and nocturnal, more so during the hottest times of the year.

**DIET.** In a study of more than 800 museum specimens, the Long-Nosed Snake was found to be a specialized predator of lizards (66%), but mammals and snake eggs made up the remainder of the diet. Of the lizards eaten, whiptails (*Aspidoscelis)* were most frequent. Larger specimens tended to take larger prey. Prey is likely killed by constriction, and it seems probable that the Long-Nosed Snake forages for lizards while they are sleeping in their burrows. One study reported that lizards compose 66.4% of the diet and mammals 26.1%.

**REPRODUCTION**. Sexual maturity is reached in males at about 498 mm and in females at 536 mm. Courtship and mating are poorly known. Females have been reported to lay clutches of up to eleven eggs (average about six) from June to August. Incubation requires 42 to 90 days, with an average of 72 days.

Figure 126-127 An almost completely black and white Long-nosed Snake. Photography JCM

**PREDATORS**. Various predators are known to feed on Long-Nosed Snakes, including large centipedes, kingsnakes, owls, hawks, armadillos, skunks, foxes, and coyotes. Defensive behavior includes releasing anal musk, biting, and sometimes bleeding from the eyes and mouth.

Long-nosed snakes are likely imperfect mimics of coral snakes, often they have red, yellow, and black in irregular bands that presumably sends an aposematic signal to potential predators.

## California Kingsnake

*Lampropeltis californiae (*Blainville, 1835) Figures 128-130.

**IDENTIFICATION.** Adults reach 1,220 mm total length, with a tail that is 11 to 13% of the total length. Hatchlings are about 300 mm. Females lay clutches of up to 24 eggs that require 49 to 83 days of incubation. The head is barely wider than the forebody. Color and pattern are variable, but the most common pattern is 21 to 44 transverse bands of yellow or white. The bands extend onto the ventral surface. (They do not in the Desert Kingsnake.) Bands are well-developed in juveniles and young adults but start to fade in old adults. A striped phase is present in southern California but is unknown in Arizona. Dorsal scales are smooth and in 23 to 25 rows at mid-body. Ventrals are 213 to 255 and subcaudals 46 to 63.

Figures 128-130. California Kingsnakes from the Green Valley area. Photography JCM.

**DISTRIBUTION.** The California Kingsnake has one of the broadest ranges of any species of kingsnakes, occurring in most habitats west of the continental divide. It does well in urbanized environments and may be frequently encountered in semi-urban and urban habitats. The species ranges from Oregon in the north and extends southward to the Mexican states of Sonora, Baja California Norte, and Baja Sur. In Arizona, it occurs in most of the state, but is absent in Santa Cruz and most of Cochise counties. It ranges from sea level to 2,164 m.

**HABITAT.** California Kingsnakes seem to be able to survive and even thrive in some human-modified environments. At least part of the reason for this was uncovered by Anguiano and Diffendorfer (2015). They investigated the spatial ecology of this snake in unfragmented and fragmented habitats with varying sizes and degrees of exposure to urban edges. They radiotracked 34 Kingsnakes for up to three years across four types of sites: interior areas of unfragmented ecological reserves, the urbanized edges of the reserves, large habitat fragments, and small habitat fragments. They found no relationship between individual snake movements and the degree of exposure to urban edges and fragmentation. The average home-range size on each site type was smaller than the smallest fragment in which snakes were tracked. Thus, the continued presence of this species in fragmented landscapes may be related to their small spatial movement patterns, home-range overlap, and ability to use urban-edge habitat.

**FORAGING AND DIET.** They are active foragers. Prey is killed by constriction. Diet includes small mammals, birds, lizards, snakes, turtles, frogs, and salamanders, as well as a variety of invertebrates. Additionally, it will eat eggs from birds and reptiles. At some locations, it may hybridize with the Desert Kingsnake, *Lampropeltis splendida.*

The California Kingsnake may be considered semi-domesticated; it has been bred extensively in captivity for color and pattern morphs by the designer snake industry.

---

## Desert Kingsnake

*Lampropeltis splendida* Baird and Girard 1853: 131. Figures 131-132

**IDENTIFICATION.** A medium-to large-bodied an average adult size range of 90–114 cm and a maximum size of 152cm. Scales are smooth, anal plate single, with midbody scale rows typically numbering 23–25. Ventral scales number 199–227 in males and 203–237 in females, while subcaudals range from 45–62 in males and 40–52 in females. The Desert Kingsnake can be distinguished from related species primarily because of its color pattern. Black or dark brown blotches with yellow lateral and dorsolateral stippling characterize the pattern. The remnant crossbands formed by this stippling yield a row of black or brown dorsal blotches or saddles, numbering 42–97. The head is black or dark brown, and the onset of the yellow dorsal patterning sometimes gives the appearance of a collar.

**DISTRIBUTION.** The Desert Kingsnake inhabits the Chihuahuan desert east of the Cochise Filter Barrier, from western Texas to extreme southeastern Arizona, north from central New Mexico in the Rio Grande River valley south to the south-central portion of the Mexican Plateau. Additionally, ecological niche modeling from Pyron and Burbrink (2009c) predict an area of habitat in northcentral Arizona as suitable which is not expected as suitable for the geographically adjacent California lineage.

**HABITAT.** Valleys, bajadas, and probably lower slopes of mountain ranges with Sonoran Desertscrub, Chihuahuan Desertscrub, Semi-desert Grassland, Great Plain's grassland, and possibly the lower edge of oak woodlands at elevations to about 1560 m.

This species does well in agriculture and is not uncommon in riparian areas, as well as low-density urban areas. It is active day and night, and there are records for every month of the year except January in the Circle, although most are found from May through September. Most of the specimens I have found in the study area have been in the late afternoon. This snake is usually found on the ground, but it climbs very well and can be found in trees.

**REPRODUCTION**. Clutch sizes range from 5 to 12 and hatchlings are found in August and September.

Figures 131-132. Desert Kingsnake. Figure 126 ( top right) is a typical Desert Kingsnake pattern. Figure 127 is a common pattern for kingsnakes in the focus area. Photography JCM

Figure 133. A sequence of photos showing a Desert Kingsnake eating a Western Diamondback Rattlesnake. The event occurred in my neighbor's yard. Photography JCM.

## Knobloch's Mountain Kingsnake

*Lampropeltis knoblochi* Taylor 1940 Figures 134-136.

**IDENTIFICATION.** Adults reach 1,114 mm in total length, with a tail about 14 to 18% of the total length. Hatchlings are 201 to 294 mm in total length. It is one of two species of tricolor kingsnakes found in Arizona. The head is about the same width as the forebody and is black with a white snout. The pattern on the body consists of black, white-to-cream, and red-to-orange transverse bands. The black bands separate the red and the white. The ventral surface is white to cream, and the red and black bands do not intrude on the belly. Dorsal scales are smooth and in 23 rows at mid-body. The black bands end at about the $5^{th}$ scale row, so that a saddle of black is bordering the white bands. (The Arizona Mountai Kingsnake has black bands extending below the fifth scale row.) This Sky Island species occurs south of the Mogollon Rim in Arizona and ranges southward into the Sierra Madre Occidental.

**HABITAT.** Oak savanna, pine, pine-oak, and mixed coniferous forest at 1,130 m to

Figures 134-135. Knobloch's Mountain Kingsnake. Photography JCM

2,881 m. However, it may use arroyos and streams to reach lower elevations in more arid semi-desert grasslands. Rock crevices and forest leaf litter are usually where they can be seen, but they will climb trees in search of prey. It can be an active day or night from March to October.

**DIET.** Lizards, but also rodents and birds. Hobbyists refer to these snakes as pyros, which is misleading because the name also refers to the Arizona Mountain Kingsnake, *L. pyromelana*. This species was removed from the synonymy of *Lampropeltis pyromelana* by Burbrink et al. (2011).

**REPRODUCTION**. occurs in the spring, and eggs have been reported in April and May. Clutch size is 3 to 6, and the eggs require 58 to 78 days of incubation, depending on the temperature.

Figure 136. Knoblock's Mountain Kingsnake. Photography JCM

# Viperidae & Pit Vipers

Here, I follow the taxonomy of Schuett et al. (2016) for Arizona rattlesnakes. Arizona has at least 15 of the 55 species of rattlesnakes. These are among the most studied of North American snakes. Vipers are divided into three subfamilies.

The Viperinae are Eastern Hemisphere snakes without heat sensing loreal pits. While the Crotalinae are pit vipers of the Eastern and Western hemisphere and have heat sensing pits. The Causinae, contains the Night Adders of subSahara Africa. And, the Azemiopinae is composed of two species found in China, Tibet, and Vietnam. Thus all New World vipers are pit vipers.

---

## Western Diamondback Rattlesnake

*Crotalus atrox,* Baird and Girard, 1853 Figures 137-139.

**IDENTIFICATION.** Adult snakes are medium-sized and have a robust body with a triangular head. Specimens from four counties in Arizona measured between 837 and 1,058 mm in length. This species shows sexual dimorphism, with males larger and having longer tails than females. Males also have fewer ventral scales. Neonates measure 210 to 360 mm. Keeled dorsal scales are in 25 rows at mid-body. A noticeable pale stripe runs diagonally from behind each eye to the corner of the mouth. The color and pattern of the western diamondback can vary widely, allowing it to blend in with its surroundings; hues can include gray, brown, olive, tan, or yellow. The dorsum is often covered in diamond-shaped blotches, which can be brown or black with lighter edges. The tail has alternating black and white rings, just anterior to the rattle, which sets this species apart from most other rattlesnakes. Although hybrids with other species do exist, identification is challenging. The Mohave Rattlesnake, *C. scutulatus,* can be difficult to separate from some Western Diamondbacks. Many Mohave Rattlesnakes have lower contrast rings on the tail (opposed to high contrast tail rings in Western Diamondbacks); many also have a green tint to their overall coloration; and many have the postocular stripe extending towards the body (opposed to the corner of the mouth in Western Diamondbacks).

**DISTRIBUTION.** The WDR extends from Arkansas to southern California and southward into Mexico (Baja, Chihuahua, and Sonora, southward to Veracruz).

**HABITATS.** In urbanized areas, such as those with golf courses in semi-desert grasslands, these snakes are often observed basking on sidewalks, crossing roads, or waiting to ambush a rodent along a curb. Their activity increases significantly during and after the monsoon season. The addition of surface water and an increase in rodent populations, facilitated by human activities, can lead to chaning the role of Western Diamondbacks from mouse trap to pest.

**IMPORTANCE.** This species is considered the most medically significant snake in its range, as it is responsible for more serious bites to humans than any other species. Ecologically, the Western Diamondback plays a crucial role by consuming various vertebrate prey, including the spiny horned lizards and pack rats, that have a significant bite force. Juveniles feed on lizards. As they grow, their diet expands to include larger prey. They sometimes scavenge carcasses.

**REPRODUCTION.** During mating season, males engage in ritualistic combat to secure access to females. Larger males are often victorious in these contests, which can be misinterpreted as a form of mating behavior by those unfamiliar with snakes. In this combat, the males perform a dance-like display where each tries to raise their head higher than the other. The courtship process is elaborate and can last days or weeks before mating

Figures 137-139. Top. A neonate Western Diamondback with an anterior stripe. Middle. An adult Western Diamondback in a defensive position. Bottom. A Western Diamondback crossing a street in Green Valley, AZ. Photography JCM

occurs, and copulation may last more than 24 hours. Males typically do not mate with the same female multiple times in one breeding season. Female Western Diamondbacks often reproduce every other year, although annual reproduction can occur. Clark et al. (2014) revealed that while multiple paternity is frequent, larger males do not necessarily father more offspring. Nonetheless, larger males tend to monopolize females for more extended periods during mating.

Rosen and Goldberg (2002) studied mature females in Arizona and found that their litter sizes ranged from 4 to 9, with an average of approximately 5.6. Adult females average a length of about 797 mm, ranging from 648 mm to 1010 mm. Neonates from wild-caught females averaged about 319.7 mm long and weighed around 19.6 g. Females have enlarged ovarian follicles between February and June, which indicates ovulation for the current year, and again from September to November, suggesting future breeding. Young are born between late July and September, with the earliest sightings recorded in July. Females may reproduce in consecutive years when prey is abundant, with about half reproducing in an average year and even more (73%) during years with high rodent populations.

---

## Klauber's Rock Rattlesnake

*Crotalus lepidus klauberi* Gloyd, 1936. Figures 140-142.

**IDENTIFICATION.** Klauber's Rock Rattlesnake can grow up to 828 mm in total length, although most adults are about 400 mm. Their tails account for 5.4% to 10.4% of their total length. Neonates typically measure 166 mm to 189 mm. adult males are larger than females, and the dorsal coloration of these snakes is sexually dimorphic; males typically have a green background, while females have a gray background. Both sexes have 13 to 24 narrow, dark brown, or black bands across their bodies and one to six bands on their tails. Occasionally, dark flecks can create incomplete bands between the primary bands. Their dorsal scales are keeled, arranged in a pattern of about 23 rows at mid-body. This color pattern helps with camouflage, especially when they rest on lichen-covered rocks. Young rattlesnakes start with a bright yellow tail that fades as they mature. In Arizona, it can be identified by its distinctive dorsal pattern and the absence of a postocular stripe, which differentiates this from other rattlesnake species.

**ACTIVITY.** Adult females tend to be found basking on sun-warmed rocks while pregnant, allowing them to maintain their body temperature. In contrast, adult males prefer cooler areas, particularly those with greenish lichen-covered rocks. Young rattlesnakes are born during the monsoon season, coinciding with the appearance of new born and newlt hatched lizards, which makes them an ideal target for feeding. Studies indicates that the diet of this snake in Arizona consists of 55.4% lizards, 28.3% centipedes, 13.8% mammals, 1.9% birds, and 0.6% snakes, with most lizards belonging to the genus *Sceloporus.*

**HABITAT.** It is typically found at elevations between 1,495 and 2,500 m, commonly between 1,675 and 2,135 m, and in Madrean Evergreen Woodland. True to its name, Klauber's Rock Rattlesnake is often found in rocky areas, particularly on large rockslides in clearings between trees. They are very cautious and tend to flee when approached, usually taking cover under nearby rocks. A patient observer can often find success by waiting at a safe distance for the snake to re-emerge after a brief period. Its habitat overlaps with other rattlesnakes, such as the Twin-Spotted and Ridge-Nosed Rattlesnakes, which also inhabit the Sky Island regions. Unfortunately, these snakes are popular among collectors, and their rocky habitats are often damaged. Klauber's Rock Rattlesnake is protected under regulations from the Arizona Game and Fish Department and collecting them requires a special

Figures 140-142. Klauber's Rock Rattlesnake. Top two are males the bottom image shows female. Photography by JCM.

permit. There is no open season for hunting this species.

Despite its small size and short fangs, this snake can inflict significant damage with its bite. Some populations in Arizona possess neurotoxic components in their venom.

## Western Black-Tailed Rattlesnake

***Crotalus*** *molossus* Baird & Girard, 1953. Figures 143-144

**IDENTIFICATION.** Adults reach 1,300 mm, but most are about 1,000 mm; tails are 4.6 to 8.6% of the total length. The neonates are 229 to 315 mm. The forebody is slender, and the head is broad and triangular. The snout is usually black or dark brown. The keeled dorsal scales are in 25 to 29 rows at mid-body. The dorsum can be brown, gray, yellow-brown, or gold-yellow with a series (22 to 39) of large, black or brown hexagonal dorsal blotches. The blotches have jagged edges, resulting in each scale being one color. The centers of the blotches are pale. Posteriorly, the blotches narrow and become transverse crossbands. The tail is black, sometimes with muted, dark, gray-brown rings. In Arizona, especially in the southeast part of the state at higher elevations, Blacktails have a honey-yellow wash, making them among the most attractive serpents in the area. At lower elevations, Blacktails are a muted green-gray with drab blotches, and the snout is not marked with dark pigment.

**DISTRIBUTION.** In the southern half of the state, from the Mogollon Rim southward, but they are absent from the western lowland deserts. They occur well into Mexico, where other subspecies replace them. Blacktails are often associated with rocks and riparian areas. Recent genetic evidence demonstrated a western form (*C. molossus*) and an eastern form(*C. ornatus*). Ironically, the species called *ornatus* is less ornate than *molossus. Crotalus ornatus,* found in New Mexico and Texas, tends to have a drabber gray coloration overall. Regardless of the name, the two forms now recognized as species were previously referred to in the literature simply as the Black-tailed Rattlesnake or the Northern Black-tailed Rattlesnake.

**HABITAT.** In Arizona, Western Black-Tails are mountain dwellers, ranging in elevation from 350 to 2,895 m in the Pinalenos. Thus, they cover various habitats, from desert foothills to montane coniferous forests. However, they seem most abundant in oak and pine-oak woodlands.

**PARENTAL CARE.** Much of what we know about this snake comes from a long-term

Figures 143-144. The Western Blacktail Rattlesnake. Top is a young individual. Below is an adult in a defensive posture. Photography by JCM.

study of natural history by Harry Greene and David Hardy in southeastern Arizona. They radio-tracked rattlesnakes for many years and documented many aspects of their natural history and reproductive biology, including natal den site selection, male-female interactions, and neonatal attendance by the female parent. Until their study, the general belief was that females abandoned their young at birth, but now we know that many pit vipers stay with the young until the first shed. Observations by Greene et al. (2002) confirmed the female attendance of her litter and sociality among rattlesnakes. They found nine litters of 2 to 6 neonates (mean = 3.9). Females departed from birth sites about 11 to 12 days after giving birth. The females moved to their summer ranges and began ambush hunting. Neonates in three litters with known birth dates shed their skins 9 to 12 days after birth. About 71% of the neonates shed skins shortly before or after they had dispersed, and all but one were at or very near the birth site.

**TAXONOMY AND SYSTEMATICS.** *Crotalus molossus* was described by Baird and Girard (1853:10) from the type locality of Fort Webster, Sta. Rita del Cobre, Grant Co., New Mexico. The holotype is USNM 485, a male, collected by J. H. Clark, date of collection unknown. Hallowell (1854:192) described *Crotalus ornatus* from the type locality between El Paso and San Antonio, along the Pecos River, Texas. Holotype, USNM 486, a skin from a female, collected by A. L. Heermann. Hallowell's name is now considered to be a valid species. Boulenger applied the name *Crotalus terrificus* in error.

## Twin-Spotted Rattlesnake

*Crotalus pricei* Van Denburgh, 1895. Figure 145.

**IDENTIFICATION.** The Twin-spotted Rattlesnake is a small pit viper that reaches a total length of 660 mm, but most individuals average about 400 mm. Their tails make up about 5.6% to 11% of their overall length. Neonates, or baby snakes, typically measure 127 mm to 203 mm long. This species is characterized by its slender body and a head that is only slightly wider than its body. Its coloration ranges from gray to pale brown and features two rows of dark brown to black spots along its body, usually between 39 and 64 spots. Some of these spots may merge to create irregular, dark transverse crossbands toward the back of the snake. The tail is also marked with similar transverse bands, while the sides of the body display lateral dark spots and a distinctive dark stripe that extends from each eye to the corner of the mouth. In some cases, the snake can exhibit a uniform dorsal color. One notable characteristic is that the basal segment of the rattle appears orange in all age groups except for neonates. The dorsal scales are keeled and arranged in about 21 rows at mid-body. This unique combination of small size, coloration, and pattern makes the Twin Spotted Rattlesnake easily distinguishable from other rattlesnake species.

**DISTRIBUTION.** Southeastern Arizona (Chiricahua, Huachuca, Pinaleno, Dos Cabezas,and Santa Rita mountains) southward into the higher elevations of Mexico. (Prival et al. 2002, Campbell and Lamar 2004). The elevational range is approximately 1,220-3,200 m (Stebbins 2003); the lower limit is approximately 1,850 m, according to Campbell and Lamar (2004). Prival et al. (2002) found this species at 2,530-2,900 meters in the Chiricahua Mountains.

**REPRODUCTION.** Female Twin-spotted Rattlesnakes typically produce litters of 3 to 9 offspring. However, like many small Sky Island rattlesnakes, they are highly sought after by illegal collectors. One researcher has dedicated over a dozen years to monitoring the population in the Chiricahua Mountains to study the impacts of poaching. His observations

Figure 145. A Twin-Spotted Rattlesnake. An Adult. Photography by JCM

reveal that the average size of female snakes decreased over the study period, likely due to poachers focusing on collecting the largest individuals. Larger females tend to produce more offspring, so this practice significantly hampers population growth.

## Mojave Rattlesnake

*Crotalus scutulatus* Kennicott, 1861 Figures 146-147.

**IDENTIFICATION.** Adults reach 1,236 mm, but most are 700 to 900 mm; tails are 5.5 to 7.7% of the total length. Neonates average about 265 mm. This is a heavy-bodied rattlesnake with a thin forebody and a postocular stripe extending from behind the eye diagonally to the back of the head, but it does not cross the labial scales. (In Western Diamondbacks, this stripe goes to the corner of the mouth.) The background coloration can vary from green, gray, yellow, tan, olive-green, to brown. The dorsal pattern is composed of irregular, dark, diamond- or near-diamond-shaped markings. The subspecies present in Arizona is the Northern Mojave Rattlesnake, *C. s. scutulatus*. Nomenclature aside, this snake is like the Western Diamondback Rattlesnake, with which it is often confused. A common vernacular term is "Mojave Green," as many populations or individuals exhibit a lime-green hue, although they may appear brown in some regions of Arizona and beyond. Adults are superficially like Western Diamondbacks but tend to have distinctly broader white bars than black bars on the "coon tail." Also, the Mojave's postocular stripe tends to extend above and well past the corner of the mouth (it extends to the corner of the mouth in the Western Diamondback), and the dorsal scales tend to be a solid brown. At the same time, those of the Western Diamondback are speckled (best seen with magnification). Defense behavior includes rattling and striking but may also involve hiding the head under a coil and exhibiting saltation (jumping forward).

**DISTRIBUTION** The name is a bit of a misnomer because it occurs in the three deserts found in Arizona, as well as the volcanic lowlands of south-central Mexico. The largest

Figure 146-147. Top. A juvenile Mojave Rattlesnake. Bottom Figure 146. An adult Mojave Rattlesnake in defensive posture. Photography JCM.

portion of its range is in the Chihuahuan Desert. Although it is found in deserts, it is also a grassland species that inhabits mountain valleys and semi-desert grasslands in southeastern Arizona. However, it was named *Mojave* because it is the only medium-to large-sized rattlesnake of the creosote bush flats in the Mojave Desert.

**DIET.** The Mojave Rattlesnake feeds on lizards, rodents, and ground-nesting birds.

**VENOM.** The Mojave Rattlesnake has a high volume of extremely potent venom, and in theory, one large individual has enough venom to kill about nine healthy adult humans. However, rattlesnakes typically only discharge a portion of their venom on a defensive strike, reserving most for feeding. The good news is that the venom from this species is used to manufacture antivenom to treat snakebite. Thus, if treated for snakebite, antitoxins of the Mojave Rattlesnakes are already in the serum. Despite its high toxicity, very few people die from bites by this or any other species, but medical care is a must. The victim may not exhibit symptoms of envenomation for some time, but the strong neurotoxins have a profound effect on the nervous system. The signal from the nerves controlling the muscles used for breathing may no longer function, so the victim may suffocate. Some populations possess the Mojave B-type toxin, which is primarily hemotoxic, affecting the blood and blood vessels. While these populations are less toxic, the venom is more destructive to tissues. One explanation for the Mojave Rattlesnake's highly neurotoxic venom (in some populations) may be attributed to kangaroo rats making up a large portion of its diet. The prey needs to be subdued quickly before it can escape by hopping away. Rattlesnakes track mammalian prey by following heat and scent trails, so an animal that makes great leaps and bounds after being struck could be difficult to find. The neurotoxic venom almost immediately paralyzes its prey.

---

## Tiger Rattlesnake,

*Crotalus tigris* Kennicott, 1859 Figures 148-149.

**IDENTIFICATION.** Adults can grow up to 959 mm long, but most individuals typically measure between 500 and 630 mm. Their tails represent about 6.4% to 8.4% of their total length, and neonates usually measure around 210 mm at birth. This rattlesnake exhibits a highly variable coloration, even among individuals from the same area. Most specimens are seen in shades of blue-gray or orange-brown. Its back is adorned with 35 to 52 transverse crossbands formed by darkly pigmented scales. The dorsal scales are keeled and arranged in about 23 rows at the snake's mid-body. The head is relatively small and triangular, and the forebody is slender, making the Tiger Rattlesnake quite distinctive among rattlesnakes.

**DISTRIBUTION.** The distribution of Tiger Rattlesnakes is interesting. In the United States, they primarily inhabit the foothills and lower mountains of the Arizona upland division of the Sonoran Desert. They have also been documented in the Peloncillo Mountains, which straddle the boundary between the Sonoran and Chihuahuan deserts and the semi-desert grasslands of the Apache Highlands. In Mexico, they can be found in similar habitats, ranging from Sonoran Desert scrub to oak woodlands. Research conducted by Goode and colleagues (2016) involved long-term monitoring of Tiger Rattlesnakes at various sites. They discovered that these snakes often den in rocky locations and remain near their dens until the arrival of the monsoon season, which usually lasts from July to September. During this time, males begin foraging in washes, and it is during the monsoon that one may have a greater chance of spotting male Tiger Rattlesnakes on the move along backroads or while hiking. Typically, they are most active in the evening, but on mild, overcast, and humid days, they may also be seen active during daylight.

**VENOM.** The Tiger Rattlesnake has venom that is slightly more toxic than that of the

Figures 148-149. The Tiger Rattlesnake individuals in both photos are adults. Photography JCM

Mojave Rattlesnake. However, its overall volume of venom is significantly lower, making it less dangerous in practical terms. Additionally, encounters with this species are uncommon, resulting in relatively few bites. Tiger Rattlesnakes are generally docile creatures, but they can and will defend themselves if threatened.

**DIET.** They primarily feed opportunistically on mammals and lizards, and their small head may assist them in accessing small prey hidden in burrows.

**REPRODUCTION.** Female Tiger Rattlesnakes typically produce small litters ranging from 1 to 6 offspring.

---

## Arizona Ridge-Nosed Rattlesnake

*Crotalus willardi* Meek, 1906 Figures 150-151.

**IDENTIFICATION.** Adults grow to 662 mm, with tails 7.5 to 11.9% of the total length. Neonates are 165 to 222 mm in total length. The species has a distinctive canthal ridge that borders the entire snout and is formed by upturned scales. The dorsum ground color is dark brown or gray with 18 to 40 dark blotches separated by pale or white transverse crossbands that may be edged with darker pigment. The most notable characteristic of this subspecies is probably the pattern of bold white and reddish-brown markings on the face. On the ventral surface, the anterior is pale, becoming darker posteriorly. Dorsal scales are keeled and in 23 to 25 rows at mid-body. Neonates are unusual because they only have a diminutive pre-button and do not develop loose rattles for several months. The tail of neonates is also black, a feature that disappears with age. Like some other species, it has been observed to attract prey with caudal luring. Adults do have a functional rattle, but these are such small snakes that they may not even be heard when rattling.

**DISTRIBUTION.** The Arizona Ridge-Nosed Rattlesnake has a very restricted distribution, occurring only in the Huachuca, Patagonia, Santa Rita, and Whetstone mountains, as well as the Canelo Hills. In Mexico, its range is also small and disjunct, only being found in five of the adjacent Mexican Sky Islands. It ranges in elevation from 1,525 to 2,750 m

**HABITAT.** While it is found in a variety of montane habitats, it appears to be most abundant along the slopes of draws and drainages with bunchgrasses or logs. It may also be found in upland situations, in rocky terrain, but it is not obligated to use that habitat like the other small montane rattlesnakes. It uses bunchgrasses, logs, and rocks as ambush areas from which it can strike its prey.

**DIET**. Unlike the other small montane rattlesnakes, adults feed primarily on small mammals, rather than lizards, but lizards are probably the most important prey of juveniles. Like Rock Rattlesnakes, they also feed on centipedes.

**VENOM**. The Arizona Ridge-Nosed Rattlesnake is regarded as one of the least dan-

Figure 150-151. An adult Arizona Ridge-Nosed Rattlesnake. Photography JCM

gerous rattlesnakes because of its small size, small fangs, and low venom yield. However, venom toxicity appears to be highly variable, and bites by this species are rarely reported. Descriptions of bite symptoms attributed to this snake range from being like a bee sting to being somewhat serious.

**CONSERVATION.** Barker (2016) suggested that each subspecies of *C. willardi* should be recognized as a full species. I am following this arrangement. This is another Sky Island rattlesnake that is often targeted by poachers. The Arizona Game and Fish Department-completely protects the species.

## The Sonoran Coral Snake

The molecular clock suggests that the American coral snake lineage split from the Asian coral snakes about 25.2 million years ago, with the ancestor dispersing from Asia to the Western Hemisphere. Today, there are approximately 86 species of coral snakes, distributed across three genera, in Asia and the Americas. Asian coral snakes (five species of *Sinomicrurus*) are the closest relatives of the two genera of Western Hemisphere coral snakes (*Micruroides* and *Micrurus*). Fossils from the European Miocene have been identified as those of coral snakes (*Micrurus*). If these are indeed coral snakes, three possible scenarios are outlined by Kelly et al. (2009):

The Sonoran Coral Snake, *Micruroides euryxanthus,* is Arizona's only coral snake species. In the history of coral snakes, it is the sister of all the other New World coral snakes. Thus, it is the closest living lineage to the ancestral coral snake, the ancestor that colonized the Western Hemisphere. Three subspecies of the Sonoran Coral Snake are recognized, but only one, the Arizona Coral Snake (*M. e. euryxanthus*), is found in Arizona. The others are restricted to Mexico's west coast.

---

### Sonoran Coral Snake

*Micruroides euryxanthus* Kennicott, 1861 Figures 152-155.

**IDENTIFICATION.** Adults grow to 660 mm, with the tail about 8% of the total length. Hatchlings are 190 to 203 mm in total length. The head is black and about the same width as the forebody. The dorsum has 11 to 13 transverse bands of red, white, and black. The red and white (yellow) bands are in contact and encircle the body; the bands on the tail are black and white. Many harmless snakes have brightly colored, banded patterns that mimic coral snakes to a greater or lesser degree. In the United States, most of the harmless mimics do not have the red and yellow bands in contact. ("Red and yellow kill a fellow.") The exceptions to this are the shovel-nosed snakes of the genus *Chionactus*, which have a color sequence identical to the coral snake pattern. However, the shovel-nosed snakes have red bands on the tail. Despite Kennicott's description of the color pattern, the red bands often have scales with some black pigment. Arizona specimens have been found with the red completely replaced by black, giving them a two-toned coloration.

**DISTRIBUTION.** Sonoran Coral Snakes are distributed southward from central Arina and southwest New Mexico to Mazatlán and Sinaloa, Mexico. They occur from 320 to 1,770m ASL in a variety of habitats, including grassland, agricultural lands, desert, brushland, thornscrub, thorn forest, tropical deciduous forest, and lower pine-oak woodland, as well as in some urban situations. The Sonoran Coral Snake in Arizona is often found in rocky bajadas and canyons, alongside upland desert scrub and grasslands. This species is perhaps most common in arroyos and river bottoms. Its fossorial habits and nocturnal activity result in infrequent encounters with humans. However, it may be locally common and is most often encountered during the monsoon season when humidity is high.

**DIET.** Like other coral snakes, the Sonoran Coral Snake feeds on elongated prey, primarily other snakes. Threadsnakes (*Rena*) are often found in their digestive systems. Females lay clutches of up to six eggs in mid-summer, with the hatchlings appearing in September.

**DEFENSE.** Defensive behaviors include hiding its head under a coil, waving its tail in the air, and cloacal popping. The venom is neurotoxic, and human envenomations have been reported; however, its small size, short fangs, and non-aggressive nature make bites relatively uncommon. Coral snakes are a minimal threat to humans and their pets. Despite their neurotoxic venom, Sonoran Coral Snakes have their share of predators, including the kingsnakes (*Lampropeltis*), the whipsnakes (Coluber), and a wide range of carnivorous birds and mammals. One unusual predator of this snake is the Black Widow, Lactrodectus hesperus. Of course, humans are also predatory on this snake, killing them because of a preceived threat.

VENOM. Many authors have proclaimed this coral snake species is not medically im-

Figure 152-154. The Sonoran Coral Snake, Photography JCM

portant. However, there have been 22 bites reported between 1955 and 2005. Symptoms of envenomation range from numbness at the bite site or on the face, to a reduction in muscular control, headache, nausea, difficulty focusing the eyes, and drooping eyelids No fatalities are known from this snake, probably the result of its small venom yield (6 mg). See Russel (1967) and Jones (2022). At this writing there is no antivenom for the Sonoran Coral Snake.

Figure 155. A Sonoran Coral Snake crossing an open area with little vegetation. Photography JCM

# References


Amarello M, Smith JJ. 2012. Social Snakes? The role of kin selection in rattlesnake aggregations. Sonoran Herpetologist 25(12):129.

Anderson JD. 1962. Egg laying and nesting in *Sceloporus scalaris slevini*. Herpetologica 18(3):162–4.

Anderson RA 1993. An analysis of foraging in the lizard *Cnemidophorus tigrinus*. Pp 84–116, In Wright and Vitt (eds.) Biology of Whiptail Lizards (Genus *Cnemidophorus*). Oklahoma Museum of Natural History.

Anguiano MP, Diffendorfer JE. 2015. Effects of fragmentation on the spatial ecology of the California Kingsnake (*Lampropeltis californiae*). Journal of Herpetology 49(3):420–7.

Archie JW, Bezy RL, Enderson EF. 2006. Yarrow's spiny lizard (*Sceloporus jarrovii* Cope 1875): Lowe's line revisited. Sonoran Herpetologist 19:50–3.

Averill-Murray RC, CM Klug 2000. Monitoring and ecology of Sonoran Desert tortoises in Arizona. Nongame and Endangered Wildlife Program Technical Report 161. Arizona Game and Fish Department, Phoenix.

Babb RD, Bradley GL, Brennan TC, Holycross AT. 2005. Preliminary assessment of the diet of *Gyalopion quadrangulare* (Serpentes: Colubridae). The Southwestern Naturalist 50(3):390-2.

Baird SF. 1858. Description of New Genera and Species of North American Lizards in the Museum of the Smithsonian Institution; Remarks on the Lower Cretaceous Beds of Kansas and Nebraska, Together with Descriptions of Some New Species of Carboniferous Fossils from the Valley of .... Proceedings of the Academy of Natural Sciences of Philadelphia. 1858:222–72.

Ballinger RE. 1974. Reproduction of the Texas horned lizard, *Phrynosoma cornutum*. Herpetologica 60: 321–327.

Barbour T. 1922. Rattlesnakes and spitting snakes. Copeia 1922:36-38.

Barker, DG, 2016. Ridgenose Rattlesnake. Pp.655-699. In: Rattlesnakes of Arizona (ed. Schuett et al. Eco Publishing, Rodeo, NM.

Barrows CW, Hoines J, Vamstad MS, Murphy-Mariscal M, Lalumiere K, Heintz J. 2016. Using citizen scientists to assess climate change shifts in desert reptile communities. Biological Conservation 195:82-8.

Bateman HL, Chung-MacCoubrey A. 2012. Growth and activity of *Sceloporus cowlesi* (southwestern fence lizard). Herpetological Review 43(1): 39–41.

Bateman HL, Chung-MacCoubrey A. 2011. *Plestiodon obsoletus* (Great Plains skink): Life history. Herpetological Review 42:276–277.

Beck DD 2005. Biology of Gila Monsters and Beaded Lizards. University of California Press, Berkeley.

Bernheim HA, Kluger MJ. 1976. Fever and antipyresis in the lizard *Dipsosaurus dorsalis*. American Journal of Physiology 231:198-203.

Berry KH, Morafka DJ, Murphy RW. 2002. Defining the desert tortoise(s): our first priority for a coherent conservation strategy. Chelonian Conservation and Biology 4:249–262.

Best TL Pfaffenberger GS. 1987. Age and sexual variation in the diet of collared lizards (*Crotaphytus collaris*). The Southwestern Naturalist. 9:415-26.

Bezy RL 1967. Variation, distribution, and taxonomic status of the Arizona night lizard (Xantusia arizonae). Copeia 1967:653—661.
Bezy RI. 2010. Elegant Earless Lizard, Holbrookia elegans Bocourt, 1874. The Sonoran Herpetologist 23(12):177-180.
Bezy RL. 2024. The Use of Scientific Names for Arizona Amphibians and Reptiles. Sonoran Herpetologists 37(2):76-83.
Bezy RL Cole CJ. 2014. Amphibians and Reptiles of the Madrean Archipelago of Arizona and New Mexico. American Museum Novitiates Number 3810:1-24.
Bezy RL, Enderson EF, Bonine KE. (2004). "Tlaconete Pinto *Pseudoeurycea bellii* (Grey, 1850) Arizona's Lost Salamander." Sonoran Herpetologist, 17, 119-122.
Blaine RA. 2008. Biogeography of the North American Southwest sand lizards. Ph.D. Dissertation. Washington University, Saint Louis, Missouri.
Blair WF. 1955. Mating call and stage of speciation in the *Microhyla olivacea-M. carolinensis* complex. Evolution, 1:469- 80.
Bogert CM. 1962. Isolation mechanisms in toads of the *Bufo debilis* group in Arizona and western Mexico. American Museum Novitates 2100:1–37.
Bogert CM, Del Campo RM. 1956. The Gila Monster and its allies. Bulletin of the American Museum of Natural History 109, 1–249.
Bohme M. Collen B, Baillie JEM, Bowles, Chanson PJ, Cox N, Hammerson G, Hoffmann M, Livingstone SR, Ram M. 2013. The conservation status of the world's reptiles. Biological Conservation 157:372–385.
Bouskila A. 1995. Interactions between predation risk and competition: a field study of kangaroo rats and snakes. Ecology. 76(1):165-78.
Bowke RW, Sullivan BK. 1991. Natural history notes: *Bufo punctatus* x *Bufo retiformis* (Red-spotted toad, Sonoran green toad). Natural hybridization. Herpetological Review 22:54.
Brattstrom BH. 1952. The food of the night lizards, genus *Xantusia.* Copeia. 1952(3):168–72.
Brennan TC, Holycross AT. 2006. A field guide to amphibians and reptiles in Arizona. Arizona Game and Fish Department.
Brown DE, editor. 1994. Biotic communities: southwestern United States and northwestern Mexico. University of Utah Press.
Bryson RW, De Oca ANM, Jaeger JR. Riddle BR. 2010. Elucidation of cryptic diversity in a widespread Nearctic treefrog reveals episodes of mitochondrial gene capture as frogs diversified across a dynamic landscape. Evolution, 64: 2315-2330.
Bryson RW, Jaeger JR, Lemos-Espinal JA, Lazcano D. 2012. A multilocus perspective on the speciation history of a North American aridland toad (*Anaxyrus punctatus*). Molecular Phylogenetics and Evolution, 64:393-400.
Bryson RW, Smith BT, De Oca ANM, García Vázquez UO, Riddle BR. 2014. The role of mitochondrial introgression in illuminating the evolutionary history of Nearctic treefrogs. Zoological Journal of the Linnean Society, 172: 103-116.
Burbrink FT, Yao H, Ingrasci M, Bryson RW, Guiher TJ, Ruane S. 2011. Speciation at the Mogollon Rim in the Arizona mountain kingsnake (*Lampropeltis pyromelana*). Molecular Phylogenetics and Evolution. 60(3):445-54.

Bury RB, Whelan JA. 1984. Ecology and management of the bullfrog. US Department of the Interior, Fish and Wildlife Service.
Campbell B. 1931. *Rana tarahumarae*, a frog new to the United States. Copeia 1931:164.
Carpenter CC, Gillingham JC, Murphy JB. The combat ritual of the rock rattlesnake (*Crotalus lepidus*). Copeia. 1976 Dec 30:764-80.
Caron, M. D., and D. E. Swann. 2009. Great Plains Skink, pp. 456-459 *in* Lizards of theAmerican southwest. A Photographic Field Guide. Rio Neuvo Publishers.
Clark RW, Schuett GW, Repp RA, Amarello M, Smith CF, Herrmann H-W. 2014. Mating Systems, Reproductive Success, and Sexual Selection in Secretive Species: A Case Study of the Western Diamond-Backed Rattlesnake, *Crotalus atrox*. PLoS ONE 9(3):
Claunch N, Taylor E. 2017. *Gambelia wislizenii* (Long-nosed Leopard Lizard) Saurocaudophagy. Herpetological Review 48:190–192.
Cole CJ, Taylor HL, Neaves WB, Baumann DP, Newton A, Schnittker R, Baumann P. 2017. The Second Known Tetraploid Species of Parthenogenetic Tetrapod (Reptilia: Squamata: Teiidae): Description, Reproduction, Comparisons with Ancestral Taxa, And Origins of Multiple Clones. Bulletin of the Museum of Comparative Zoology. 161(8):285-321.
Conlon JM, Mechkarska M, Coquet L, Jouenne T, Leprince J, Vaudry H, Kolodziejek J, Nowotny N, King JD. Characterization of antimicrobial peptides in skin secretions from discrete populations of *Lithobates chiricahuensis* (Ranidae) from central and southern Arizona. Peptides. 2011 Apr 30;32(4):664-9.
Cooper JG. 1861. New Californian animals. Proceedings of the California Academy of Sciences (ser. 1) 2:118–123.
Corl A, Davis AR, Kuchta SR, Comendant T, Sinervo B. 2010. Alternative mating strategies and the evolution of sexual size dimorphism in the side-blotched lizard, *Uta stansburiana*: A population-level comparative analysis. Evolution. 64(1):79-96.
Coues E. 1875. Synopsis of the Reptiles and Batrachians of Arizona; with critical and field notes, and an extensive synonymy. Chapter IV. Reports upon the Collections Obtained from Portions of Nevada, Utah, California, Colorado, New Mexico, and Arizona, during the Report upon Geographical and Geological Explorations and Surveys West of the One Hundredth Meridian. Volume V, Zoology: 509-633.
Crother BI. (ed.). 2017. Scientific and standard English names of amphibians and reptiles of North America north of Mexico, with comments regarding confidence in our understanding, 8th ed. SSAR Herpetological Circular 37:1–101.
Cruz-Sáenz D, Lazcano D, Ontiveros-Esqueda L, Narváez-Torres S, Flores-Covarrubias E. Notes on the Herpetofauna of Western Mexico 4: A Case of Parental Care in the Mountain Skink (*Plestiodon callicephalus*) in Huaxtla, Zapopan, Jalisco, Mexico. Bulletin of the Chicago Herpetological Society 46:57–61.
Degenhardt, W.G., C.W. Painter, and A.H. Price. 1996. Amphibians & Reptiles of New Mexico. University of New Mexico Press. Albuquerque, New Mexico.
DeNardo DF, Moeller KT, Seward M, Repp R. 2018. Evidence for atypical nest

overwintering by hatchling lizards, *Heloderma suspectum*. Proceedings of the Royal Society Biology 285(1879):20180632.

Dimmitt MA, Ruibal R. 1980. Exploitation of food resources by spadefoot toads (*Scaphiopus*). Copeia. 1980:854-62.

Dinehart SK, Smith LM, McMurry ST, Smith PN, Anderson TA & Haukos DA. 2010. Acute and chronic toxicity of Roundup Weathermax® and Ignite® 280 SL to larval *Spea multiplicata* and *S. bombifrons* from the Southern High Plains, USA. Environmental Pollution, 158:2610-2617.

Dittmer DE, Johnson JB, Hibbitts TJ. 2015. Sexual Dimorphism and Patch Size Variation in Three Lizard Species Suggests Potential for Sexual Confusion. Copeia. 2015:310–21.

Dodd CK. 2013. Frogs of the United States and Canada, 2-vol. John Hopkins University Press, Baltimore.

Douglas ME, Douglas MR, Schuett GW, Porras L, Holycross A. 2002. Phylogeography of the western rattlesnake (*Crotalus viridis*) complex, with emphasis on the Colorado Plateau. Pp 11-50 In Schuett, Hoggren, Douglas, and Green (eds.) Biology of the Vipers, Eagle Mountain Publishing.

Duellman WE, Marion AB, Hedges B. 2016. Phylogenetics, classification, and biogeography of the treefrogs (Amphibia: Anura: Arboreanae). Zootaxa. 4104:1–109.

Dunn ER. (1926). The Salamanders of the Family Plethodontidae. Smith College, Northhampton, Massachusetts. Eifler DA, Eifler MA, Brown TK. 2012. Habitat selection by foraging Texas horned lizards, *Phrynosoma cornutum*. The Southwestern Naturalist 57:39–43.

Echternach AT. 1967.Ecological Relationships of Two Species of the Lizard Genus *Cnemidophorus* in the Santa Rita Mountains of Arizona. The American Midland Naturalist, 78:448-459.

Emerson JD . 2024. *Trimorphodon lambda* (Sonoran Lyre Snake). ENVENOMATION. Herpetological Review, 55(1):126-127.

Ennen JR, Lovich JE, Averill-Murray RC, Yackulic CB, Agha M, Loughran C, Tennant L, Sinervo B. 2017. The evolution of different maternal investment strategies in two closely related desert vertebrates. Ecology and Evolution 7:3177–89.

Ernst CH, Ernst EM. 2003. Snakes of the United States and Canada. Smithsonian Books, Washington, D.C. Ernst CH, Lovich JE. 2009. Turtles of the United States and Canada. John Hopkins University Press, Baltimore.

Figueroa A, McKelvy AD, Grismer LL, Bell CD, Lailvaux SP. 2016. A Species-Level Phylogeny of Extant Snakes with Description of a New Colubrid Subfamily and Genus. PloS One. 2016 11(9):e0161070.

Fisher MC, Garner TW, Walker SF. 2009. Global emergence of *Batrachochytrium dendrobatidis* and amphibian chytridiomycosis in space, time, and host. Annual Review of Microbiology, 63:291-310.

Fisher MC, Ghosh P, Shelton JM, Bates K, Brookes L, Wierzbicki C, Rosa GM, Farrer RA, Aanensen DM, Alvarado-Rybak M, Bataille A. 2018. Development and worldwide use of non-lethal, and minimal population-level impact, protocols for the isolation of amphibian chytrid fungi. Scientific Reports 8(1):777.

Fleischner, T. L., Floyd, M. L., Rack, J., Hanna, D., Blevins, K., Christman, B., and Holycross, A. T. 2024. The Mogollon Highlands Ecoregion of the American Southwest: A Neglected Center of Ecological Diversity. Natural Areas Journal, 44(2):104-119.

Fontanella FM, Feldman CR, Siddall ME, Burbrink FT. 2008. Phylogeography of *Diadophis punctatus*: extensive lineage diversity and repeated patterns of historical demography in a trans-continental snake. Molecular Phylogenetics and Evolution 46:1049–70.
Fontanella FM, Miles E, Strott P. 2021. Integrated analysis of the ringneck snake *Diadophis punctatus* complex (Colubridae: Dipsadidae) in a biodiversity hotspot provides the foundation for conservation reassessment. Biological Journal of the Linnean Society 133(1):105-19.
Foote R, MacMahon JA. 1977. Electrophoretic studies of rattlesnake (*Crotalus* & *Sistrurus*) venom: taxonomic implications. Comparative Biochemistry and Physiology Part B: Comparative Biochemistry. 57(3):235-41.
Fowlie, JA. 1965. The Snakes of Arizona. Azul Quinta Press.
Gámez N, Nihei SS, Scheinvar E, Morrone JJ. 2016. A temporally dynamic approach for cladistic biogeography and the processes underlying the biogeographic patterns of North American deserts. Journal of Zoological Systematics and Evolutionary Research. doi: 10.1111/jzs.12142.
Germano DJ. 2014. Activity, Growth, Reproduction, and Population Structure of Desert Box Turtles (*Terrapene ornata luteola*) at the Northern Edge of the Chihuahuan Desert. Chelonian Conservation and Biology. 2014:56–64.
Germano DJ, Hungerford CR. 1981. Reptile population changes with manipulation of Sonoran Desert shrub. The Great Basin Naturalist. 1981:129-38.
Gillingham JC, Baker RE. 1981. Evidence for scavenging behavior in the western diamondback rattlesnake (*Crotalus atrox*). Ethology. 55(3):217-27.
Glaudas X, Goldberg SR, Hamilton BT. 2009. Timing of reproduction of a cold desert viperid snake from North America, the Great Basin rattlesnake (*Crotalus lutosus*). Journal of arid environments. 73(8):719-25.
Glenn JL, Straight RC. 1977. The midget faded rattlesnake (*Crotalus viridis concolor*) venom: Lethal toxicity and individual variability. Toxicon 15: 129–133.
Goldberg C. 2003. Barking Frog (*Eleutherodactylus augusti*). The Tucson Herpetological Society. Sonoran Herpetologist 16(7)2003.
Goldberg CS, Schwalbe CR. 2000. Population ecology of the barking frog. Final Report to Arizona Game and Fish Department, Phoenix, Arizona. IIPAM project no. I98014. 50 pp.
Goldberg CS, Sullivan BK, Malone JH, Schwalbe CR, 2004. Divergence among barking frogs (*Eleutherodactylus augusti*) in the southwestern United States. Herpetologica 60:312–320.
Goldberg SR, Mahrdt CR. 2011. Reproduction in the Great Basin Collared Lizard, *Crotaphytus bicinctores* (Squamata: Crotaphytidae). Bulletin, Southern California Academy of Sciences, 110(3):189–192.
Goldberg SR. 2004. Reproduction in the Western Terrestrial Garter Snake, *Thamnophis elegans* (Serpentes: Colubridae) from Arizona. Journal of the Arizona-Nevada Academy of Science 36, no. 2 (2004):77–80.
Goldberg SR. 2014. Notes on Reproduction of Plateau Fence Lizards, *Sceloporus tristichus* (Squamata: Phrynosomatidae). Bulletin Chicago Herpetological Society 49(1–4):30–9.
Goldberg SR. 2002. Reproduction in the Chihuahuan hook-nosed snake, *Gyalopion canum*, (Serpentes: Colubridae). *Plestiodon obsoletus* (Great Plains skink):

Life history. Journal of the Arizona-Nevada Academy of Sciences 34:95–97.
Goode, M, Parker MR, Feldner MJ, Repp RA. 2016. Tiger Rattlesnake. Pp. 607-654. In: Schuett et al. (eds.) Arizona Rattlesnakes. Eco Publishing.
Gottscho AD, Marks SB, Jennings WB. 2014. Speciation, population structure, and demographic history of the Mojave Fringe-toed Lizard *(Uma scoparia)*, a species of conservation concern. Ecology and evolution, 4(12), 2546–2562.
Grey JE. 1850. Catalogue of the Specimens of Amphibia in the Collection of the British Museum. Part II. Batrachia Gradientia, etc. Printed by Order of the Trustees. Spottiswoodes and Shaw, London.
Greene HS. 1997. Snakes, the evolution of mystery in nature. University of California Press, Berkeley.
Gutiérrez-González CE, Gómez-Ramírez MA, Gutierrez-Garcia D, Valenzuela J, Rorabaugh JC, Flesch AD. 2016. Interactions between Sonoran Desert Toads (*Incilius alvarius*) and Mammalian Predators at the Northern Jaguar Reserve, Sonora, Mexico. 2016. Sonoran Herpetologist 29(2):26.
Hahn, DE 1979. *Leptotyphlops humilis* (Baird and Girard) Western Blind Snake. Catalogue of American Amphibians and Reptiles 232, 1–4.
Hallowell E. 1853. On a new genus and three new species of reptiles inhabiting North America. Proceedings of the Academy of Natural Sciences of Philadelphia 6: 206–209.
Hardy LM. 1975. A Systematic Revision of the Colubrid Snake Genus *Gyalopion*. Journal of Herpetology 9:107–132.
Hekkala ER., Saumure RA, Jaeger JR, Herrmann H-W, Sredl MJ, Bradford DF, Drabeck, Blum MJ. 2011. Resurrecting an extinct species: archival DNA, taxonomy, and conservation of the Vegas Valley leopard frog. Conservation Genetics 12, no. 5 (2011): 1379–1385.
Howell PE, Hossack BR, Muths E, Sigafus BH, Chandler RB. 2016. Survival estimates for reintroduced populations of the Chiricahua leopard frog (*Lithobates chiricahuensis*). Copeia. 104(4):824-30.
Holm PA. 2008. Phylogenetic Biology of the Burrowing Snake Tribe Sonorini (Colubridae). ProQuest;
Holman JA. 2000. Fossil Snakes of North America: Origin, Evolution, Distribution, Paleoecology. Indiana University Press, Bloomington.
Holycross AT, Jones LLC. In Review. *Crotalus lepidus*: Rock Rattlesnake (Kennicott 1861). In: Holycross AT, Mitchell JC (eds.). Snakes of Arizona. Eco Publishing.
Holycross AT, Painter CW, Prival DB, Swann DE, Schroff MJ, Edwards T, Schwalbe CR. 2002. Diet of *Crotalus lepidus klauberi* (banded rock rattlesnake). Journal of Herpetology. 36(4):589-97.
Holycross AT, Schwalbe CR. 1995. *Lampropeltis triangulum*. Geographic Distribution. Herpetological Review 26:46.
Hossack BR, Honeycutt RK, Sigafus BH, Muths E, Crawford CL, Jones TR, Sorensen JA, Rorabaugh JC, Chambert T. 2017. Informing recovery in a human-transformed landscape: Drought-mediated coexistence alters population trends of an imperiled salamander and invasive predators. Biological Conservation 209:377-94.

Husak JF, Fox SF. 2003. Adult male collared lizards, *Crotaphytus collaris*, increase aggression towards displaced neighbours. Animal Behaviour. 65(2):391–6.

Husak JF, Macedonia JM, Fox SF, Sauceda RC. 2006. Predation cost of conspicuous male coloration in collared lizards (*Crotaphytus collaris*): an experimental test using clay-covered model lizards. Ethology 112:572–80.

Irschick DJ, Shaffer HB. 1997. 'The polytypic species revisited: Morphological differentiation among tiger salamanders (*Ambystoma tigrinum*) (Amphibia: Caudata). Herpetologica, 53(1), 30–49.

Jackson K, Kley NJ, Brainerd EL. 2004. How snakes eat snakes: the biomechanical challenges of ophiophagy for the California kingsnake, *Lampropeltis getula californiae* (Serpentes: Colubridae). Zoology. 107(3):191–200.

Jacob JS, Altenbach JS. 1977. Sexual Color Dimorphism in *Crotalus lepidus klauberi* Gloyd (Reptilia, Serpentes, Viperidae). Journal of Herpetology. 11:81-4.

Jani AJ, Briggs CJ. Host and aquatic environment shape the amphibian skin microbiome, but effects on downstream resistance to the pathogen Batrachochytrium dendrobatidis are variable. Frontiers in Microbiology 2018;9:487.

Jaeger JR., Riddle BR., Jennings RD, Bradford DF. 2001. Rediscovering *Rana onca*: evidence for phylogenetically distinct leopard frogs from the border region of Nevada, Utah, and Arizona. Copeia, 2001(2), 339–354.

Jaeger JR, Waddle AW, Rivera R, Harrison DT, Ellison S, Forrest MJ, Vredenburg VT, van Breukelen F. 2017. *Batrachochytrium dendrobatidis* and the Decline and Survival of the Relict Leopard Frog. EcoHealth. 14(2):285-95.

Jennings MR. 1988. Rana onca. Pp. 417.1–417.2. Catalogue of American Amphibians and Reptiles. Society for the Study of Amphibians and Reptiles, St. Louis, Missouri.

Jennings MR., Hayes MP. 1994. Amphibian and Reptile Species of Special Concern in California. Inland Fisheries Division, California Department of Fish and Game, Rancho Cordova, CA.

Jones B. 1971. Influence of prey on the feeding behavior of *Trimorphodon biscutatus lambda* (Colubridae). The Southwestern Naturalist, 33(4), 488-490.

Jones KB, Kepner LP, Martin TE. 1985. Species of reptiles occupying habitat islands in western Arizona: a deterministic assemblage. Oecologia. 66(4):595-601.

Jones KB. 1988. Distribution and habitat associations of herpetofauna in Arizona: comparisons by habitat type. General technical report RM-Rocky Mountain Forest and Range Experiment Station, US Department of Agriculture, Forest Service (USA).

Jones LL, Lovich RE. 2009. Lizards of the American Southwest: a photographic field guide. Rio Nuevo Publishers, Tucson.

Jones LLC. 2016. The spelling of Mojave vs. Mohave as it applies to Standard English names of reptiles and amphibians. Sonoran Herpetologist 29 (4):65-70.

Jones LLC. 2022.. Venomous Animals of the American Southwest. Tucson. Rio Nuevo Publishers.

Jones LLCJ, Halama KJ, Lovich RE (eds). 2016. Habitat Management Guidelines for Amphibians and Reptiles of the Southwestern United States. Partners in Amphibian and Reptile Conservation, Technical Publication HMG- 5, Birmingham, AL. 193 pp.

Jones TR, Collins JP, Kocher TD, Mitton JB. 1988. Systematic status and distribution of *Ambystoma tigrinum stebbinsi* Lowe (Amphibia: Caudata). Copeia. 1988:621–35.

Kelly CM, Barker NP, Villet MH, Broadley DG. 2009. Phylogeny, biogeography and classification of the snake superfamily Elapoidea: a rapid radiation in the late Eocene. Cladistics 25(1):38—63.

Kennicott R. 1860 (1861). Descriptions of new species of North American serpents in the museum of the Smithsonian Institution, Washington. Proceedings of the Academy of Natural Sciences Philadelphia 12: 328—338.

Klauber LM. 1939. A new subspecies of the western worm snake. Transactions of the San Diego Society of Natural History 9 (14):67-68.

Knox SC, Chambers C, Germaine SS. 2001, Habitat Associations of the Sagebrush Lizard (*Sceloporus graciosus*): Potential Responses of an Ectotherm to Ponderosa Pine Forest Restoration Treatments. Proceedings. RMRS-P-22. 2001.

Koludarov I, Jackson TN, Sunagar K, Nouwens A, Hendrikx I, Fry BG. 2014. Fossilized venom: the unusually conserved venom profiles of *Heloderma* species (beaded lizards and gila monsters). Toxins. 6:3582–95.

Krysko KL, Nuñez LP, Newman CE, Bowen BW. 2017. Phylogenetics of Kingsnakes, *Lampropeltis getula* Complex (Serpentes: Colubridae), in Eastern North America. Journal of Heredity 2017, 1:1–13.

Kwiatkowski MA, Sullivan BK. 2002. Mating system structure and population density in a polygynous lizard, *Sauromalus obesus*. Behavioral Ecology 13:201-8.

Lannoo M. (ed). 2005. Amphibian Declines: The Conservation Status of United States Species. University of California Press.

Lappin AK, Swinney EJ. 1999. Sexual dimorphism as it relates to natural history of leopard lizards (Crotaphytidae: *Gambelia*). Copeia. 1999:649–60.

Lappin AK., Husak JF. 2005. Weapon performance, not size, determines mating success and potential reproductive output in the collared lizard (*Crotaphytus collaris*). The American Naturalist, 166:426–436.

Lattanzio MS, Metro KJ, Miles DB 2014 Preference for Male Traits Differ in Two Female Morphs of the Tree Lizard, *Urosaurus ornatus*. PLoS ONE 9(7): e101515.oi:10.1371/journal.pone.0101515

Leaché AD, Linkem CW. 2015. Phylogenomics of Horned Lizards (Genus: *Phrynosoma*) Using Targeted Sequence Capture Data. Copeia, 103(3), 586–594.

Leydekker R. 1896. The Royal Natural History Vol. 5. London Frederick Warne & Co.

Levis NA, Serna Buzón S, Pfennig DW. 2015. An inducible offense: carnivore morph tadpoles induced by tadpole carnivory. Ecology and evolution, 5(7), 1405-1411.

Lovich JE, LaRue CT, Drost CA, Arundel TR. 2014. Traditional cultural use as a tool for inferring biogeography and provenance: a case study involving painted turtles (*Chrysemys picta*) and Hopi Native American culture in Arizona, USA. Copeia. 2014:215–20.

Lowe C. 1955. The Salamanders of Arizona. Transactions of the Kansas Academy of Science, 58: 237–251.

Lowe CH, Jones CJ, and Wright J W. 1968. 'A new plethodontid from Sonora, Mexico.' Contributions in Science, Natural History Museum of Los Angeles County, 411, 11-22.

McGuire J. 1996. Phylogenetic systematics of crotaphytid lizards (Reptilia: Iguania: Crotaphytidae). Bulletin of the Carnegie Museum of Natural History 32:1–143.

McGuire JA, Linkem CW, Koo MS, Hutchison DW, Lappin AK, Orange DI, Lemos-Espinal J, Riddle BR, Jaeger JR. 2007. Mitochondrial introgression and incomplete lineage sorting through space and time: phylogenetics of crotaphytid lizards. Evolution. 61:2879–97.

Meik JM. 2016. Southwestern Speckled Rattlesnake. Pp. 531-562. in: Schuett et al. (eds.) Arizona Rattlesnakes. Eco Publishing.

Mendelson JR, Mulcahy DG, Williams TS, Sites Jr JW. 2011. A phylogeny and evolutionary natural history of Mesoamerican toads (Anura: Bufonidae: *Incilius*) based on morphology, life history, and molecular data. Zootaxa 3138:1-34.

Meshaka Jr WE, Collins SL, Bury RB, McCallum ML. 2022. Exotic amphibians and reptiles of the United States. University Press of Florida.

Meyers JJ, Herrel A, Nishikawa KC. 2006. Morphological correlates of ant eating in horned lizards (*Phrynosoma*). Biological Journal of the Linnean Society. 89(1):13–24.

Mitchell JC. 1979. Ecology of southeastern Arizona whiptail lizards (*Cnemidophorus*: Teiidae): population densities, resource partitioning, and niche overlap. Canadian Journal of Zoology. 57(7):1487–99.

Mitchell JC, Holycross AT, Rosen PC. 2005. Phyllorhynchus browni, Saddled Leaf-Nosed Snake. Pages 287-292 in Holycross and Mitchell (eds.) Snakes of Arizona. Rodeo, New Mexico, Eco Publishing.

Montanucci RR. 1978. Dorsal pattern polymorphism and adaptation in *Gambelia wislizenii* (Reptilia, Lacertilia, Iguanidae). Journal of Herpetology, 73–81.

Mulcahy DG. 2008. Phylogeography and species boundaries of the western North American Nightsnake (*Hypsiglena torquata*): revisiting the subspecies concept. Molecular phylogenetics and evolution. 46(3):1095–1115.

Murphy RW, Berry KH, Edwards T, Leviton AE, Lathrop A, Riedle JD. 2011. The dazed and confused identity of Agassiz's land tortoise, *Gopherus agassizii* (Testudines, Testudinidae) with the description of a new species, and its consequences for conservation. ZooKeys 113: 39–71.

Musgrave ME, Cochran DM. 1929. *Bufo alvarius*, a poisonous toad. Copeia, 1929: 96-99.

Myers EA, Burgoon JL, Ray JM, Martínez-Gómez JE, Matías-Ferrer N, Mulcahy DG, Burbrink FT. 2017. Coalescent Species Tree Inference of *Coluber* and *Masticophis*. Copeia. 2017 105(4):642-50.

Nieuwolt-Dacanay PM. 1997. Reproduction in the western box turtle, *Terrapene ornata luteola*. Copeia. 1997 Dec 9:819–26.

Noonan BP, Pramuk JB, Bezy RL, Sinclair EA, de Queiroz K, Sites JW. 2013. Phylogenetic relationships within the lizard clade Xantusiidae: using trees and divergence times to address evolutionary questions at multiple levels. Molecular phylogenetics and evolution. 69(1):109–22.

Norris KS, Kavanau JL. 1966. The burrowing of the Western Shovel-nosed Snake, *Chionactis occipitalis* Hallowell, and the under- s a n d environment. Copeia (4):650–664.

Norris KS. 1953. The ecology of the desert iguana *Dipsosaurus dorsalis*. Ecology

34(2):265-87.

Nowak, EM, Persons TB. 2010. Milksnakes at Petrified Forest National Park, Arizona: adaptive monitoring or rare vertebrates. Pages 133–149 in The Colorado Plateau IV: Shaping conservation through science and management. Van Riper, Wakeling and Sisk (eds.) University of Arizona Press.

O'Connell KA, Smith EN. 2018. The effect of missing data on coalescent species delimitation and a taxonomic revision of whipsnakes (Colubridae: *Masticophis*). Molecular phylogenetics and evolution. Mar 20.

O'Donnell RP, Mixan R, Leavitt DJ, Ingraldi MF, Branch WC. 2016. Surveys for threatened and endangered species and their potential breeding habitat: Yellow-billed Cuckoos and Mexican Garter Snakes of the big sandy and Santa Maria Rivers, Colorado River district, Arizona. Arizona Game and Fish Department, Wildlife Contracts Branch, Phoenix, Arizona, USA.

Papenfuss TJ, Macey JR, Schulte JA. 2001. A new lizard species in the genus *Xantusia* from Arizona. Natural History Museum, University of Kansas, 23:1–9.

Parker JM, Anderson SH. 2007. Ecology and Behavior of the Midget Faded Rattlesnake (*Crotalus oreganus concolor*) in Wyoming. Journal of Herpetology. 41(1):41-51.

Parker WS. 1972. Aspects of the ecology of a Sonoran Desert population of the western banded gecko, *Coleonyx variegatus* (Sauria, Eublepharinae). American Midland Naturalist 1:209–24.

Parra-Olea G, Garcia-Paris M, Papenfuss TJ, Wake DB. 2005. Systematics of the *Pseudoeurycea bellii* (Caudata: Plethodontidae) species complex. Herpetologica, 61, 145-158.

Pfennig DW, Murphy PJ. 2002. How fluctuating competition and phenotypic plasticity mediate species divergence. Evolution, 56(6), 1217-1228.

Pfennig KS. 2007. Facultative mate choice drives adaptive hybridization. Science 318 965-967. Pianka ER, Parker WS. 1975. Ecology of horned lizards: a review with special reference to *Phrynosoma platyrhinos*. Copeia, 141–162.

Piotrowski JS, Annis SL, Longcore JE. 2004. Physiology of *Batrachochytrium dendrobatidis*, a chytrid pathogen of amphibians. Mycologia 96:9–15.

Pitts SL, Hughes BD, Mali I. 2017. Rattlesnake Nuisance Removals and Urban Expansion in Phoenix, Arizona. Western North American Naturalist. 77(3):309-16.

Place AJ, Abramson CI. 2004. A quantitative analysis of the ancestral area of rattlesnakes. Journal of Herpetology 38:152-6.

Platz JE, Clarkson RW, Rorabaugh JC, Hillis DM. 1990. *Rana berlandieri*: Recently introduced populations in Arizona and southeastern California. Copeia 1990(2):324-333.

Plummer MV. 2003. Activity and thermal ecology of the Box Turtle, *Terrapene ornata*, at its southwestern range limit in Arizona. Chelonian Conservation and Biology 4:569–577.

Plummer MV. 2004. Seasonal inactivity of the desert box turtle, *Terrapene ornata luteola*, at the species' southwestern range limit in Arizona. Journal of Herpetology. 38:589–93.

Pook CE, Wüster W, and Thorpe RS. 2000. Historical biogeography of the western rattlesnake (Serpentes: Viperidae: *Crotalus viridis*), inferred from mitochondrial DNA sequence information. Molecular Phylogenetics and Evolution 15:269-282.

Prival DB. 2007. The short life and times of a twin-spotted rattlesnake. Sonoran Herpetologist 20(10):104-109.

Prival DB, Porter JP. 2016. Rock Rattlesnake. Pp. 432-459. in: Schuett et al. (eds.) Arizona Rattlesnakes. Eco Publishing.

Punzo F. 1974. Comparative analysis of the feeding habits of two species of Arizona blind snakes, *Leptotyphlops h. humilis* and *Leptotyphlops d. dulcis*. Journal of Herpetology. 29:153–6.

Pyron RA, Burbrink FT. 2009. Neogene diversification and taxonomic stability in the snake tribe Lampropeltini (Serpentes: Colubridae). Molecular Phylogenetics and Evolution 52(2):524–9.

Radke WR, Malcom JW. 2009. Notes on the ecology of Green Ratsnakes (*Senticolis triaspis*) in southeastern Arizona. Herpetological Conservation and Biology. 4(1):9–13.

Recuero E, Martinez-Solano I, Parra-Olea G, Barcia-Paris M. 2006. Phylogeography of *Pseudacris regilla* (Anura: Hylidae) in western North America, with a proposal for a new taxonomic rearrangement. Molecular Phylogenetics and Evolution 39:293-304.

Reiserer RS, Schuett GW. 2008. Aggressive mimicry in neonates of the sidewinder rattlesnake, *Crotalus cerastes* (Serpentes: Viperidae): stimulus control and visual perception of prey luring. Biological Journal of the Linnean Society. 95(1):81-91.

Reiserer RS, Schuett GS. 2016. The Origin and Evolution of the Rattlesnake Rattle: Misdirection, Clarification, Theory, and Progress. Chapter 7, in: Rattlesnakes of Arizona, Vol. 2.

Reiserer RS, Schuett GW, Greene HW. 2018 Seed ingestion and germination in rattlesnakes: overlooked agents of rescue and secondary dispersal. Proceedings of the Royal Society B: Biological Sciences 285: 20172755. http://dx.doi.org/10.1098/rspb.2017.2755.

Rodda G. 2020. Lizards of the World. John Hopkins, Baltimore.

Rorabaugh J. 2004. Barking Frogs (*Eleutherodactylus augusti*) of the Santa Rita Mountains. Sonoran Herpetologist. 2004;17(7):72.

Rorabaugh JC. 2010. Conservation of amphibians and reptiles in northwestern Sonora and southwestern Arizona. Pages 181-204 in: W. Halvorson, C. Schwalbe, and C. van Riper III (editors). Southwestern Desert Resources. University of Arizona Press, Tucson.

Rorabaugh JC, Espinal JL. 2016. A Field Guide to the Amphibians and Reptiles of Sonora. Eco H erpetological Publishing.

Rorabaugh JC, Hale SF, Sredl MJ, Ivanyi C. 2005. Return of the Tarahumara frog to Arizona. Connecting Mountain Islands and Desert Seas: Biodiversity and Management of the Madrean Archipelago II. Proceedings. Gottfried, GJ, BS Gebow, LG Eskew, and CB Edminster (Compilers). RMRS-P-36, US Department of Agriculture, Forest Service, Rocky Mountain Research Station, Fort Collins, Colorado, USA. 2005:345-8.

Rosen, P.C. 2003. Avra Valley snakes: Marana survey report for Tucson Shovel-nosed Snake (*Chionactis occipitalis klauberi*). Prepared for the Town of Marana, University of Arizona.

Rosen PC, Goldberg SR. 2002. Female reproduction in the western diamond-backed rattlesnake, *Crotalus atrox* (Serpentes: Viperidae), from Arizona.

Texas Journal of Science 54(4):347-56.

Rosen PC, Schwalbe CR. 1988. Status of the Mexican and narrow-headed gartersnakes (*Thamnophis eques megalops* and *Thamnophis rufipunctatus rufipunctatus*) in Arizona. Unpublished report from Arizona Game and Fish Department (Phoenix, Arizona) to U.S. Fish and Wildlife Service, Albuquerque, New Mexico.

Russell, F. E. 1967. Bites by the Sonoran coral snake Micruroides euryxanthus.Toxicon 5:39-42.

Roth-Monzón, A.J., Devitt, T.J., Ríos-Muñoz, C.A., Parra-Olea, G., Campbell, J.A. and Flores-Villela, O., 2021. Geographic Variation in the Green Rat Snake *Senticolis triaspis* (Squamata: Colubridae): Evidence from Mitochondrial DNA, Morphology, and Niche Modeling. *Ichthyology & Herpetology*, *109*(4), pp.1026-1035.

Ruthven AG. 1907. A Collection of Reptiles & Amphibians from Southern New Mexico & Arizona. Order of the Trustees, American Museum of Natural History.

Saha A, McRae L, Dodd CK, Gadsden H, Hare KM, Lukoschek V, Bohm M. 2018. Tracking Global Population Trends: Population Time-Series Data and a Living Planet Index for Reptiles. Journal of Herpetology 52(3):259-268.

Santos–Barrera G, Hammerson G. 2004. *Lithobates yavapaiensis*. The IUCN Red List of Threatened Species 2004: e.T19181A8848786.

Santoyo-Brito E, Anderson M, Fox S. 2017. Incubation Temperature Modifies Sex Ratio of Hatchlings in Collared Lizards, *Crotaphytus collaris*. Journal of Herpetology 51: 197–201.

Schlaepfer MA, Sredl MJ, Rosen PC, Ryan MJ. 2007. High prevalence of *Batrachochytrium dendrobatidis* in wild populations of lowland leopard frogs *Rana yavapaiensis* in Arizona. EcoHealth. 4(4):421.

Schuett GW, Feldner MJ, Smith CF, Reiserer RS (Eds). 2016. Rattlesnakes of Arizona. 2 Vol. Eco Publishing.

Seigel RA, Ford NB, Mahrt LA. 2000. Ecology of an aquatic snake (*Thamnophis marcianus*) in a desert environment: implications of early timing of birth and geographic variation in reproduction. The American Midland Naturalist 143(2):453–62.

Sinervo B, Mendez-De-La-Cruz F, Miles DB, Heulin B, Bastiaans E, Villagrán-Santa Cruz M, Lara-Resendiz R, Martínez- Méndez N, Calderón-Espinosa ML, Meza-Lázaro RN, Gadsden H. 2010. Erosion of lizard diversity by climate change and altered thermal niches. Science. 328(5980):894-9.

Stebbins RC. 2003. A field guide to western reptiles and amphibians. Houghton Mifflin Harcourt.

Stocking, S. 2023. *Trimorphodon lambda* (Sonoran Lyresnake). Herpetological Review 54(3):406.

Streicher JW, Cox CL, Campbell JA, Smith EN, de Sa RO. 2012 Rapid range expansion in the Great Plains narrow-mouthed toad (*Gastrophryne olivacea*) and a revised taxonomy for North American microhylids. Molecular Phylogenetics and Evolution 64:645–653.

Streicher JW. 2012. Evolutionary relationships in some northern groups of the direct-developing frog genus *Craugastor* (Anura: Craugastoridae). The University of Texas at Arlington PhD dissertation.

Strimple PD, Tomassoni AJ, Otten EJ, Bahner D. 1997. Report on envenomation by a Gila monster (*Heloderma suspectum*) with a discussion of venom apparatus, clinical findings, and treatment. Wilderness & environmental medicine. 8(2):111–6.

Sug DW, Fitzgerald LA, Snell HL. 1995. Growth rate, timing of reproduction, and size dimorphism in the southwestern earless lizard (*Cophosaurus texanus scitulus*). The Southwestern Naturalist, 193–202.

Sullivan BK and G. Rice. 2021. *Urosaurus ornatus* (Ornate Tree Lizard). Predators. Herpetological Review 52(3):655.

Sullivan BK, Sullivan E, 2016. *Gopherus morafkai* (Sonoran Desert Tortoise) Drinking behavior. Herpetological Review 47 (1):123-124.

Sullivan BK, Weinstein SA. 2017. Foraging and prey handling in the Western Patch-nosed Snake (*Salvadora hexalepis*), and consideration of venom use in non-front-fanged colubroid snakes. Herpetological Review 48:19–23.

Sullivan BK. Vernon JK. 2015. *Dipsosaurus dorsalis* (desert iguana) urban habitats. Herpetological Review 46 (1): 90-91 Tennessen JA, Zamudio KR. 2008. Genetic d i fferentiation among mountain island populations of the striped plateau lizard, *Sceloporus virgatus* (Squamata: Phrynosomatidae). Copeia. 2008 (3):558–64.

Tinkle DW. 1961. Population structure and reproduction in the lizard *Uta stansburiana stejnegeri*. American Midland Naturalist. 206–34.

Tinkle DW. 1967. The life and demography of the side-blotched lizard, *Uta stansburiana*. University of Michigan Museum of Zoology: Miscellaneous Publications (132).

Tinkle DW, McGregor D, Dana S. 1962. Home range ecology of *Uta stansburiana stejnegeri*. Ecology. 43(2):223–9.

Townsend TM, Mulcahy DG, Noonan BP, Sites JW, Kuczynski CA, Wiens JJ, Reeder TW. 2011. Phylogeny of iguanian lizards inferred from 29 nuclear loci, and a comparison of concatenated and species-tree approaches for an ancient, rapid radiation. Molecular Phylogenetics and Evolution. 61:363–80.

Van Denburgh J. 1897. The reptiles of the Pacific coast and Great Basin: An account of the species known to inhabit California, and Oregon, Washington, Idaho and Nevada. California Academy of Sciences.

Van Devender TR. 1995. Desert grassland history: changing climates, evolution, biogeography, and community dynamics. In: The Desert Grassland (Eds. McClaran MP, Van Devender TR), pp. 68–99. University of Arizona Press, Tucson.

Van Devender TR. 2000. The deep history of the Sonoran Desert. In: A Natural History of the Sonoran Desert (Eds. Phillips SJ, Comus PW), pp. 61–69. Arizona-Sonora Desert Museum Press, Tucson.

Van Devender TR, Van Devender W. 1975. Ecological notes on two Mexican skinks (genus *Eumeces*). The Southwestern Naturalist 20:279–82.

Wake DB. 2017. Persistent Plethodontid Themes: Species, Phylogenies, and Biogeography. Herpetologica. 73(3):242- 51.

Walde AD, Currylow A, Walde AM. 2015. Discovery of a new burrow associated with the Desert Tortoise (*Gopherus agassizii*), the Long-nosed Leopard Lizard (*Gambelia wislizenii*). Herpetology Notes. 108:107-9.

Walker JM, Sullivan BK, Cordes JE. 2021. Sanctioned Nomenclature for Triploid

Parthenogenetic Colorado Checkered Whiptail and for Other Taxa of the Lizard Genus *Aspidoscelis* (Family Teiidae). Herpetological Review ;52(3):550-558.

Weil AT, Davis W. 1994. *Bufo alvarius*: a potent hallucinogen of animal origin. Journal of Ethnopharmacology, 41:1-8.

Werler JE, Dixon JR. 2000. Texas Snakes: Identification, Distribution, and Natural History. Univ. Texas Press, Austin.

Wilgenbusch, J. and de Queiroz, K. 2000. Phylogenetic relationships among the phrynosomatid sand lizards inferred from mitochondrial DNA sequences generated by heterogeneous evolutionary processes. Systematic Biology 49: 592-612

Wilson LD, Mata-Silva V. 2015. A checklist and key to the snakes of the *Tantilla* clade (Squamata: Colubridae), with comments on taxonomy, distribution, and conservation. Mesoamerican Herpetology. 2:418-98.

Wood DA, Fisher RN, Vandergast AG. 2014. Frayed boundaries: color and gene flow patterns among parapatric subspecies of the Western Shovel-nosed Snake (*Chionactis occipitalis*). PLOS ONE 9: 1489–1507.

Wright AH, Wright AA. 1949. Handbook of frogs and toads of the United States and Canada. Comstock Publishing Associates. Ithaca, New York. pp. 366–368.

Wright JW, Lowe CH. 1968. Weeds, polyploids, parthenogenesis, and the geographical and ecological distribution of all-female species of *Cnemidophorus*. Copeia. 1968, 128–38.

Youssef MK, Adolph SC, Richmond JQ. 2008. Evolutionarily conserved thermal biology across continents: the North American lizard *Plestiodon gilberti* (Scincidae) compared to Asian Plestiodon. Journal of Thermal Biology 33:308–312.

Zweifel RG 1956. A survey of the frogs of the *augusti* group, genus *Eleutherodactylus*. American Museum Novitates 1813:1–35.

Zweifel RG. 1962. Notes on the distribution and reproduction of the lizard *Eumeces callicephalus*. Herpetologica 18:63– 65.

Zweifel RG 1967. Amphibia: Aura: Leptodactylidae *Eleutherodactylus augusti*. Catalogue of American Amphibians and Reptiles. 41.1–41.4.

# Glossary

Ambient Temperature - The temperature of the environment surrounding the animal in question.

Amniote – reptiles, birds , and mammals that have three embryonic membranes: the chorion, allantois, and the yolk sac.

Anamniote – fishes and amphibians have eggs that lack the three membranes found in reptiles, birds, and mammals.

Aquatic – Living in water.

Arboreal - Tree-dwelling.

Aestivation - Dormancy during summer or dry season.

Allele - Any of the different gene versions that may occur at a specific location on a chromosome..

Amelanistic – The absence of melanin, the black pigment.

Amino acid - The main building blocks of proteins, which are bonded in many ways to create many different proteins.

Amplexus - The sexual embrace of a male frog or toad: the clasping of the female's body around the back by the male's forelimbs until eggs are laid and fertilized.

Anerythristic - Lacking the red pigment.

Anterior - located near or toward the head.

Antivenom - An antidote to venom, usually from a snake, that is given to a snakebite victim. Creating anti-venom involves injecting venom into horses, rabbits, or other animals, then extracting their blood to create a serum. Antivenom for coral snake bites is not the same as that used for pit viper bites. It is sometimes called "antivenin."

Anuran, Anura – frogs and toads.

Apical Pits - Tiny depressions on the posterior portions of the dorsal scales in some snakes.

Attenuated – Thin, slender, or elongated.

Axanthic – The absence of yellow pigment.

Barbels - Small, fleshy, usually downward projections of the skin on the chin and/or throat (in some kinds of turtles and the tadpoles of the Mexican Burrowing Toad).

Basking / Sunning - An animal exposing itsself to absorb heat from the sun or objects in the environment

*Batrachochytrium dendrobatidis* (*Bd*) A disease of amphibians (frogs and toads) that damages the skin and causes the death of the amphibians.

Binomial Nomenclature - The system of scientific classification and identification of species in which the genus and an epithet are combined to make a species name. Example: *Homo sapiens* or *Crotalus atrox.*

Biodiversity - The numbers, distribution, and abundance of species within a given area.

Bioindicator - A species whose health, condition, or presence at the individual level, or the population level, is indicative of the condition of the habitat or ecosystem.

Biomass - The weight of living things in the environment.

Boss - A raised rounded knob on top of the head between the eyes or on or near the end of the snout (on some toads).

Bridge - The connection between the carapace and plastron of a turtle's shell.

Canthus Rostralis - The area between the eye and the nostril. It may be a ridge, a smooth surface, concave surface, or a convex service.

Carapace- Upper part of a turtle's shell.

Caudal luring - An ambush strategy using its tail to draw prey into striking distance.

Chromosome - A long string of DNA that is wrapped around proteins so it can be folded into a compact package. Chromosomes are usually paired in animals, and humans have 23 pairs of chromosomes.

Cladistics - a method of classification of animals and plants according to the proportion of measurable characteristics that they have in common. It is assumed that the higher the proportion of characteristics that two organisms share, the more recently they diverged from a common ancestor.

Clade- a group of organisms descended from a common ancestor. The ancestor and its descendants compose the clade.

Class - In the scientific classification system (taxonomy), a class is the division between phylum and order.

Cline - A gradual change in a variable characteristic.

Cloaca - The chamber into which the digestive, urinary, and reproductive systems empty, opening to the outside.

Codominant - Genes that, when present together, will both influence the appearance of the same trait.

Codon - A sequence of 3 base pairs that codes for either a specific amino acid, or the end of a protein, known as a "stop codon."

Cold-Blooded - An animal whose metabolism and body temperature are dependent on external heat sources. Also called ectothermy.

Communal - A community or group of several individuals.

Concertina locomotion - A method of locomotion used by snakes in which they anchor the posterior portion of their body, extend the anterior portion, and then pull themselves forward.

Copulation and Copulate - To engage in sexual intercourse or mating.

Costal grooves - Vertical grooves on a salamander's side.

Courtship - The process that precedes mating in reptiles and usually involves actions such as rubbing or biting by the male to increase the receptivity of a female to mating.

Cranial crests - Bony ridges on the head of some toads - interorbital (between the eyes) or postorbital (behind the eyes).

Crepuscular - Active at twilight and/or dawn.

Cryptic—It can refer to coloration and pattern that blend an animal with its environment. It is also used to make one species look like another.

Cryptic coloration - A type of camouflage that makes creatures difficult to spot against their background.

Deceptive coloration is when an organism's color and pattern fool its predators or prey. There are two types of deceptive coloration: camouflage and mimicry.

Defecation - The process of eliminating waste through the cloaca.

Desiccate – Remove water from organisms to dehydrate or dry out completely.

Dewlap - A vertical, loose flap of skin on the throat of some iguanid lizards. Some of them have muscles attached to display a color and pattern.

Dimorphism - Difference in form, color, or structure between members of the same species. The sexes may be different, or there may be two color phases of the same sex (dichromatism).

Diurnal - Active during daytime hours.

DNA - The large molecule that carries the instructions on how to build an organism and the substances with which the genetic code is built. The DNA molecule is a component of codons, genes, and chromosomes.

Dorsal - About the back or upper surface of the body.

Dorsolateral - Pertaining to the area at the juncture of the back and the side.

Dorsum - The entire upper surface of an animal.

Duvernoy's Gland - A venom gland in snakes located above the teeth in the upper jaw. This gland produces secretions that flow down the anterior groove of the rear fangs under low pressure when the snake bites its prey.

Ecdysis - Shedding or molting the old skin after the new skin develops underneath.

Ecosystem - A community of living organisms (biotic factors) interacting with their physical environment (abiotic factors) within a specific area.

Ectotherm - A "cold-blooded" animal; an animal that regulates its temperature behaviorally through outside sources of heat.

Egg Tooth - A small protuberance on the beak or nose of vertebrates that are hatched from eggs, i.e., birds and reptiles. Some lizards and snakes develop a true tooth that is shed after use; birds and other reptiles generally develop an analogous epidermal horn that is reabsorbed or falls off.

Embryo - The earliest stage of a developing fertilized egg that eventually turns into offspring. This term particularly describes the time of the most rapid growth.

Endemic - Confined to, or indigenous in, a certain area or region.

Envenomate - The act of injecting venom.

Enzyme - Proteins that catalyze specific chemical reactions in living cells.

Epiglottis - A lid-like flap of fibrocartilage tissue covered with a mucus membrane, attached to the root of the tongue. It projects obliquely upwards behind the tongue and the hyoid bone. It also prevents food from entering the trachea and directs it to the esophagus.

Estivation - A state of inactivity during prolonged periods of drought or high temperatures.

Extinct - A species that is no longer in existence.

Extirpated- refers to eliminating a species from a particular region, local extinction.

Extralimital - Occurring outside the geographical area.

Family - In the scientific classification system (taxonomy), family is the division between order and genus. In theory, all members of a family shared a common ancestor.

Fauna - The animals living in a specified region.

Femoral pores - A series of openings to glands on the underside of the thighs of some lizards. They produce chemicals that send a message to other lizards.

Fertility - The ability to produce offspring.

Fertilization - The fusion of male and female reproductive cells, leading to the development of a new individual.

Fossorial - Adapted for digging, burrowing, and living in the soil.

Gene - The chemical unit of heredity responsible for producing specific traits in the developing offspring. A string of base pairs (DNA) that is the smallest unit of genetic inheritance.

Genetics - The study of heredity.

Genus - A group of closely related species that share a common ancestor. Interbreeding between organisms within the same category can occur.

Genotype - An animal's genetic makeup, usually referring to the absence or presence of specific genes of interest.

Genus - In the scientific classification system (taxonomy), genus is the division between family and species.

Gestation - The development period of young (pregnancy) and/or eggs inside the body.

Gills - External organs located on the sides of the neck and used in underwater respiration.

Granular scales - Tiny grainlike or pebblelike scales that do not overlap one another.

Gravid - Bearing eggs or developing young; pregnant.

Gular - On or about the throat. Also, the most anterior scutes (or scute) on a turtle's plastron.

Hemipenis (pl. hemipenes) - One of the paired copulatory organs of male snakes and lizards.

Heterozygous - When both corresponding genes for one locus are different versions.

Hibernation - Dormancy during winter.

Heredity - The genetic transmission of a particular trait from parent to offspring.

Herpetofauna - The species of amphibians and reptiles that inhabit a given area.

Herpetologist - A scientist who studies snakes, other reptiles, and amphibians.

Herpetology - The study of reptiles and amphibians.

Herps - An abbreviated word referring to reptiles and amphibians.

Hibernation - The dormant state of reduced metabolism to which certain animals succumb during winter months.

Homozygous - When both corresponding genes for one locus are the same version.

"Hot" - a term used when referring to venomous reptiles.

Husbandry - The management or care of living things.

Hydration - The absorption of water by the body.

Hybrids - Offspring of two different varieties, races, species, or genera.

Hypomelanism - A reduction of the black pigment.

Impaction - Inablity of the eggs to correctly move through the reproductive tract. The overloading of an internal organ, such as the intestine.

Incipient - In the first stages or beginning. An incipient population is still small but beginning to reproduce and become established.

Incubation - The process through which eggs are heated until the offspring hatch.

Indigenous - A plant, animal, or people native to a region.

Infertility - The inablity to produce offspring.

Infertile Eggs - Eggs containing no developing offspring.

Intergrades - Animals of related and adjoining subspecies that may resemble either form or exhibit a combination of their characteristics.

Interorbital - Between the eyes.

Jacobson's Organ / Vomeronasal Organ (VNO) - Snakes and lizards use their tongue to pick up particles from the air and ground and put them into this organ. The organ comprises two tiny cavities in the roof of the mouth and is lined with sensitive tissue that aids in the sense of smell. Here, the particles are identified, analyzed, and acted upon. Since these cavities are split apart from each other, the tongue itself also has to split. This is why snakes and lizards have forked tongues.

Juvenile - A young animal: prior to becoming an adult.

Keel - A ridge on individual dorsalscales of some snakes: longitudinal ridge on the carapace or plastron of turtles; the raised edge along the upper edge of the tail in some salamanders.

Kingdom - In the scientific system of classification (taxonomy), kingdom is the first division by which all known life is divided. There is some argument regarding the division of single-celled organisms, but a common five-kingdom system is often used including Animalia, Plantae, Fungi, Protista, and Monera.

Labial - Of or pertaining to the lip. Scales that border the mouth.

Labial Tooth Rows - Rows of tiny, horny projections on the lips of tadpoles.

Larva(e) - A post-hatching immature stage that differs in appearance from the adult and must metamorphose before assuming adult characters (e.g., a tadpole).

Lateral - Pertaining to the side.

Lateral Undulation - A method of locomotion used by snakes in which they use waves of the body to push against the ground's surface and move the body forward.

Lethargy - Lethargic - Drowsy, dull, lifeless, or weak.

Leucism - Leucistic - A condition where there's a partial loss of pigmentation in an animal, resulting in white, pale, or patchy coloration of the skin, fur, feathers, or scales, but not the eyes. It's a genetic condition that can affect various animals, including birds, mammals, and reptiles. Leukoderma / Leukopathia / Piebald / Vitilingo - Partial or total loss of skin pigmentation, often occurring in patches.

Locus - A specific "slot" on a chromosome corresponding to a certain location, like an address to a particular house. Different genes in the same locus can control one or more traits. A locus is generally called the gene "for" a specific trait.

Mental gland - An organ on the chin of male lungless salamanders that produces a secretion sexually stimulating to females.

Metabolism - The physical and chemical processes involved in the breakdown of food into energy, and the development of stores of chemicals within the body for use during growth and development of tissues.

Metamorphose - A period of transformation from larval to adult form (e.g., the transition of a tadpole to a frog).

Metabolic Rate - The rate at which food ingested is broken down into energy for use by the body.

Middorsal - Pertaining to the center of the back.

Midventral - Pertaining to the center of the belly.

Mimicry - A condition in which an animal looks or acts like something else, often a more dangerous animal.

Miscarriage - The loss of offspring from the body of the mother before they can survive independently.

Molecular Clock- A molecular clock is a method used to estimate the time of evolutionary events by analyzing the genetic differences between species. It's based on the assumption that mutations accumulate in DNA at a roughly constant rate over time. By comparing the DNA sequences of different species, scientists can estimate how long ago they diverged from a common ancestor.

Monotypic - The only representative of its group, such as a genus with only one species.

Mucous Membrane - The passages connecting to the outside openings in which the lining is made up of cells that produce mucus, the oral cavity, the air passages, the upper digestive tract, the rectum, etc.

Mutation - An alteration of one or more genes. Generally, these are interesting when they cause a change in the phenotype of an animal.

Naso-labial groove - A hairline groove running from the nostril to the edge of the upper lip in lungless salamanders.

Necrosis - Necrotic - Death of an individual or groups of cells, or of specific areas of tissue. Another term, "cheesy", is used to describe cell death in which the tissues resemble cottage cheese.

Neonate - Newborn.

Neoteny - A condition in which salamanders fail to metamorphose but become sexually mature and reproduce while retaining larval characteristics.

Neotenic - Mature and capable of reproduction but retaining the larval form, appearance, and habits.

Nictitating membrane - A clear third eyelid in some animals that can be drawn across the eye for protection and to moisten the eye while also keeping visibility.

Nocturnal - Active at night.

Non-venomous - An animal that, because of its delivery system or the components of its venom, is not toxic to humans.

Ocelli - Round, eyelike spots.

Ocular – Pertaining to the eye.

Oocyte - An egg that has not yet reached full development.

Opaque / Opacity – pertaining to the ability not to transmit light.

Opisthoglypha – Rear fangs -snakes which have some of the posterior maxillary teeth enlarged and grooved to aid in venom delivery. This type of venom delivery occurs in colubrid snakes.

Order - In the scientific classification system (taxonomy), order is the division between class and family.

Oviparous - Producing young by means of eggs that hatch after laying ( e.g. turtles).

Ovoviviparous - Producing young by means of membranous eggs retained within the body of the female until hatching (e.g. " live-bearing" reptiles).

Ovulation - The formation and release of the ovum (egg cell) prior to fertilization.

Parasite - An organism that lives on or in another organism at whose expense it receives nourishment.

Paravertebral Stripe - A stripe lying to one side but paralleling the mid-line of the back.

Parotid gland - A large poison-producing gland on each side of the neck or behind the eyes of some amphibians.

Parthenogenesis – Asexual reproduction by the development of an unfertilized egg ( e.g., species of whiptail lizards that reproduce only one sex-females--and--reproduce by means of unfertilized eggs.) This process is widespread in squamates.

Phalanges - The bones of the toes. (Singular: phalanx.)

Phenotype - The appearance of an organism depends on the interaction of the environment and the genotype.

Pheromone - A chemical released by an animal and used as a signal to other animals of the same species. Female snakes often release pheromones to attract male snakes during the mating season.

Phylogeny - The evolutionary relationships among different groups and species of animals.

Phylum - In the scientific classification system (taxonomy), phylum is the division between kingdom and class.

Photoperiod - The amount of daylight hours versus the night hours the animal experiences.

Pigment – molecule that gives color to a plant or animal cells.

Pit organ - Pit organs on the faces of pit vipers and some boas and pythons detect heat (infrared radiation) that is given off by warm bodies. These warm bodies include prey, threatening attackers, and even a warm place to hibernate.

Passive Integrated Transponder tag (PIT tag) is a radio-frequency identification tag about the size of a grain of rice. Each tag has a unique code that can be read with a portable scanner. PIT tags are usually implanted in the body cavity with a modified syringe.

Plastron - The lower part of a turtle's shell.

Posterior - Located at or toward the rear end of the body.

Postocular - Behind the eye.

Prehensile - Adapted for grasping or wrapping around.

Preocular - Anterior to the eye.

Prey - An animal that is caught by another for food.

Protective coloration - Coloration or color pattern of an animal that affords it protection from observation either by its predators or by its prey.

Protein - A chain of amino acids, usually one that has been folded. Proteins include enzymes that build and digest proteins. Other proteins form structures – keratin forms nails, claws, and hair.

Proteolysis - The directed degradation (digestion) of proteins by cellular enzymes called proteases or by intramolecular digestion.

Proteroglyphous - A term generally referring to a group of venomous snakes which have the front of the jaw positioned, fixed fangs that are specifically grooved for venom delivery. This type of dentition is found in elapid snakes.

Race - Subspecies.

Radiotelemetry - A radio transmitter attached to, or implanted in, an animal to track its movements by locating it using a directional antenna and radio receiver.

Rear-fanged - enlarged teeth on the middle to back of the maxillary bone.

Rectilinear locomotion - A method of locomotion used by snakes in which they crawl in a straight line and use their ribs and belly scales to push themselves forward. Rectilinear locomotion is generally used by large, slow-moving snakes when they are not in a hurry.

Regurgitate – Reverse peristalsis, partially digested stomach contents are lost by way of the mouth. Equals vomiting.

Reticulation - A network of lines.

Rejuvenate - To replenish or return to a normal state.

Rookery - A site associated with the development or care of the young.

Rugose - Wrinkled or warty.

Scute - A large scale; horny shields or plates covering a turtle's shell.

Serrate - Having projections in the shape of the teeth of a saw. Jaws of some tadpoles may be serrated.

Siblings - Offspring of the same parents, but not necessarily born at the same time.

Sibling Species - Two or more species presumably derived from a common ancestor. They often resemble one another closely and may occur together or replace each other geographically.

Solenoglypha - A term commonly refers to the venomous snakes with tubular fangs positioned at the front of the maxillary bone, namely the vipers. These fangs work like hypodermic needles, allowing for direct injection of venom.

sp. - An abbreviation for "species" when the genus but not the species is known.

Sperm - The male reproductive cells that fertilize an egg during breeding.

Spiracle - A tubelike external opening on the left side in most tadpoles to exit respiratory water.

Species - individual organisms that share significant similarities in appearance, anatomy, physiology, and genetics because they have shared a relatively recent common ancestors.

Spermatophore - A cone-shaped jellylike mass topped with a sperm capsule, deposited by male salamanders during courtship. The cloacal lips of the female pick up the sperm cap.

spp. - (plural) An abbreviation if you find more than one of a genus sp..

Stress - the biological changes that occur in the body due to adverse external influences, such as the need to defend against injury or damage.

Subcaudals - The scales beneath the tail; in a double row in most snakes, but in a single row in others. Sometimes shortened to "caudals."

Subocular - Beneath the eye.

Supraocular - Above the eye.

Supraorbital Semicircle - A row of small scales separating the supraoculars from the median head scutes in certain lizards.

Suture - A seam; the boundary between scales or scutes.

Sympatric refers to two or more populations occupying identical or broadly overlapping geographic areas.

Sympatry - the coexistence of organisms in overlapping geographical areas without interbreeding.

Tadpole - The larva of an anuran.

Taxon - The technical term for a unit or category, i.e., for a species, subspecies, genus, family, etc. (Plural: taxa.)

Taxonomy - The science of the classification of animals or plants.

Thermal Receptor - Heat-sensitive organs on either side of the head that look like small pits. These pits contain membranes sensitive to infrared radiation and allow snakes to locate their prey based on temperature differences with their environment.

Tibia - The lower leg (of toads and frogs) from heel to knee.

Trait - A characteristic that can be inherited, such as a certain distinct appearance.

Transformation - See metamorphosis.

Tubercle - A small knob or a rounded protuberance in the skin.

Tympanum - The eardrum.

Venomous - Possessing glands for synthesizing venom and the ability to deliver it in such a way that it is active (as opposed to passive).

Vent - opening of the cloaca to the outside of the body.

Venter - The entire undersurface of an animal.

Ventral - Pertaining to the abdominal or belly surface.

Ventrals – The scales covering the belly of a lizard or a snake.

Vestigial - a structure that functioned in a previous evolutionary stage or species development stage.

Vocal sac - An expandable pouch on the throat of male frogs and toads that becomes filled with air and acts as a resonating chamber when they vocalize during courtship; the sac collapses at the end of the call.

Warm-Blooded - A body that can maintain a uniform temperature without depending on external heat sources. Also describes Endothermy

Xanthic - Having a yellow or yellowish color.

Xeric - Dry.

# Appendices 1-5

## Appendix 1.

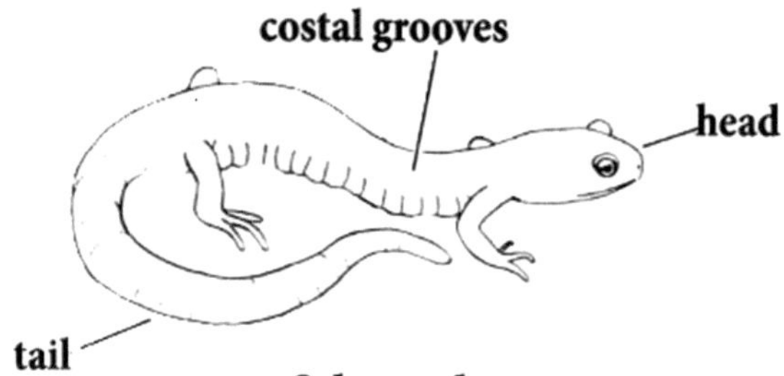

Salamander

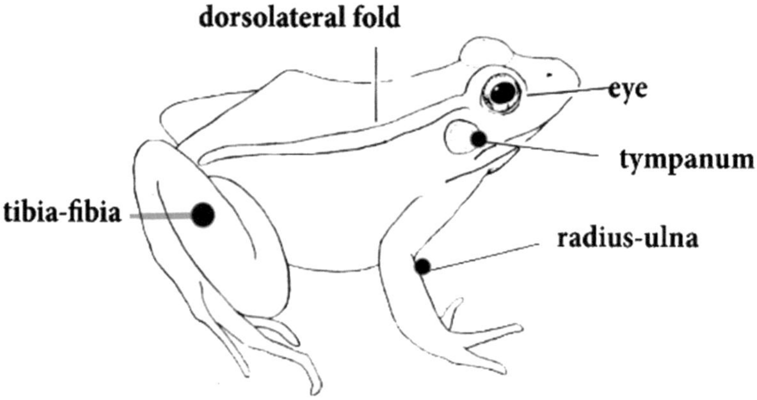

Frog

# Appendix 2.

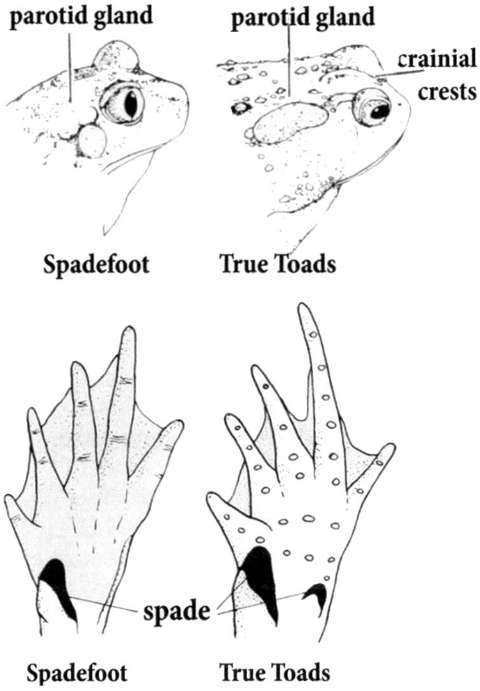
parotid gland
parotid gland
crainial
crests
Spadefoot
True Toads
spade
Spadefoot
True Toads

# Appendix 3

## Turtle Shell Nomenclature

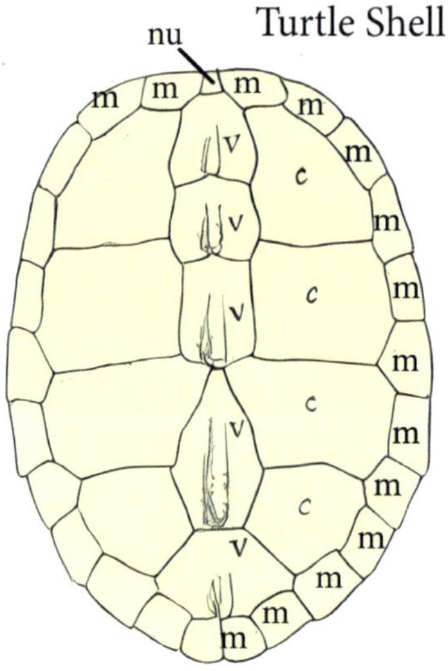


Carapace

Carapace shield terminology
nu = nuchal
v = vertebral
c =costal
m = marginals

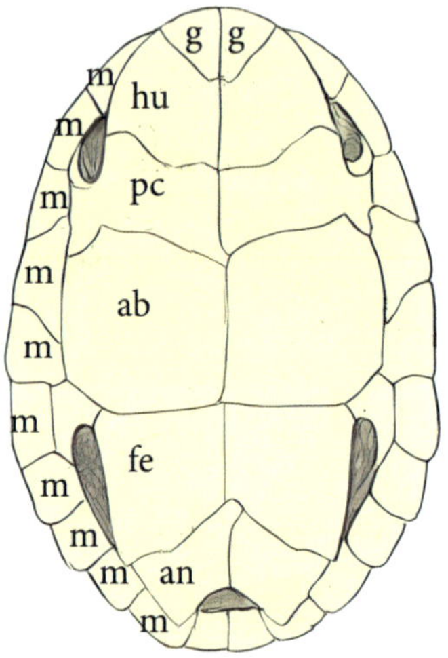


Plastron

Plastron shield terminology
g = gulars
hu = humerals
pc = pectorals
ab = abdominals
fe = femorals
an = anals

Adapted from Leydekker R. 1896. Volume V.

# Appendix 4

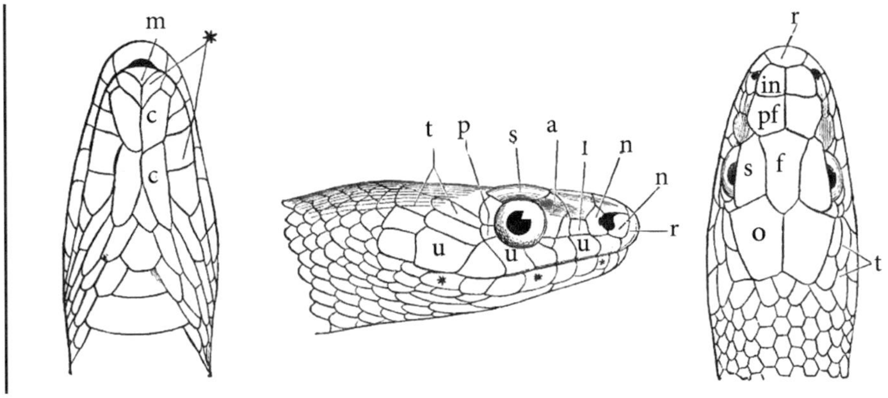


Head scale nomenclature for a snake.

Views of a snake's head. Lower surface (left), profile (middle), dorsal view (right).

Abbreviations

r = rostral; f = frontal; s = supraocular; o= parietal; n = nasal; l = loreal; a = preocular; p = postocular; u = upper labial; t = temporal; m = mental; * = lower labial; c = chin shield. Adapted from Leydekker R. 1896. Volume V.

# Appendix 5

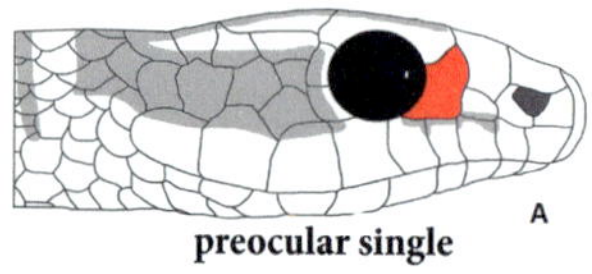


**preocular single**

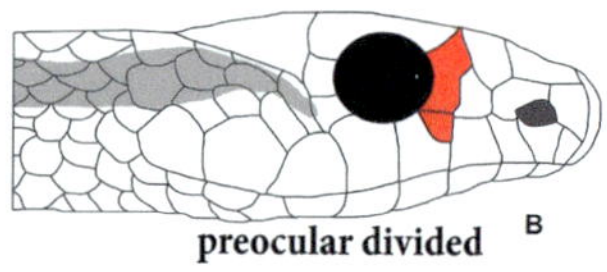


**preocular divided**

---

**a second pair of chinshields in contact with each other or separated by one row of scales**

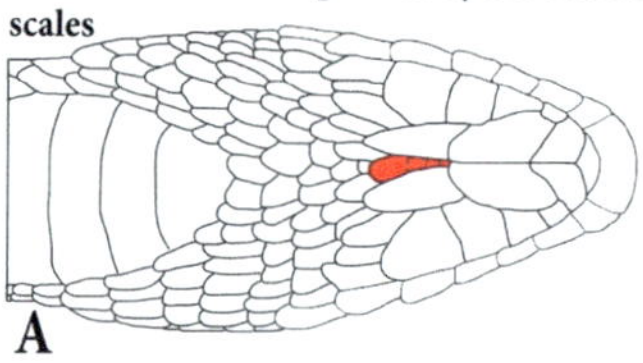


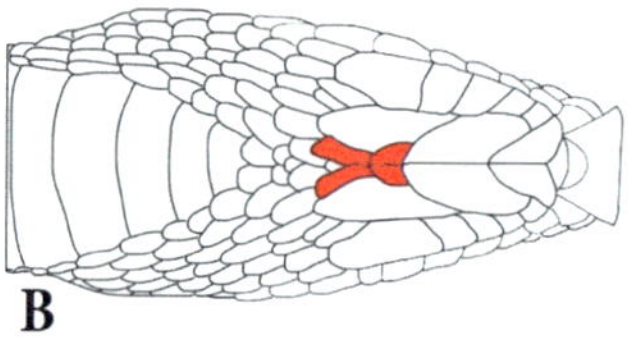


---

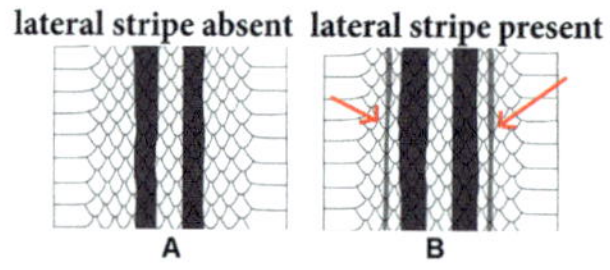


---

**two upper labials enter orbit**

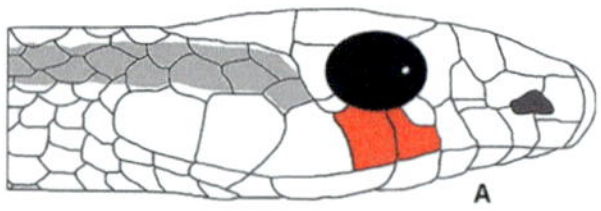


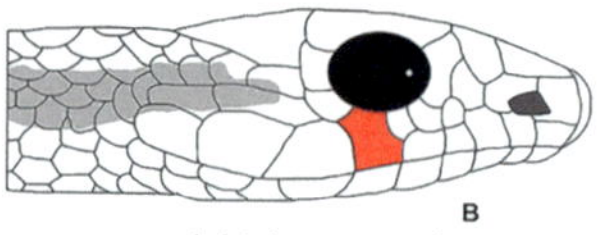


**one upper labial enters orbit**

A snake does not usually eat for a week or so before it sheds its skin. Here is a case of a snake shedding its skin (species uncertain) while it was investigating eggs in quail nest. Photography by Diane Thornton.

Made in the USA
Las Vegas, NV
03 November 2026

57424060R00112